SpringerBriefs in Earth Sciences

Series Editor

Pedro Maciel de Paula Garcia, Universidade Federal de Mato Grosso, Cuiabá, Brazil

SpringerBriefs in Earth Sciences present concise summaries of cutting-edge research and practical applications in all research areas across earth sciences. It publishes peer-reviewed monographs under the editorial supervision of an international advisory board with the aim to publish 8 to 12 weeks after acceptance. Featuring compact volumes of 50 to 125 pages (approx. 20,000–70,000 words), the series covers a range of content from professional to academic such as:

- timely reports of state-of-the art analytical techniques
- bridges between new research results
- snapshots of hot and/or emerging topics
- literature reviews
- in-depth case studies

Briefs will be published as part of Springer's eBook collection, with millions of users worldwide. In addition, Briefs will be available for individual print and electronic purchase. Briefs are characterized by fast, global electronic dissemination, standard publishing contracts, easy-to-use manuscript preparation and formatting guidelines, and expedited production schedules.

Both solicited and unsolicited manuscripts are considered for publication in this series.

Kachikwulu Kingsley Okeke

Palynofacies and Petroleum Migration Style of Inland Anambra Basin Nigeria

Unravelling Kerogen Maturation
and Structural Traps Kinematics

Kachikwulu Kingsley Okeke
Department of Geology
University of Nigeria
Nsukka, Enugu State, Nigeria

ISSN 2191-5369   ISSN 2191-5377 (electronic)
SpringerBriefs in Earth Sciences
ISBN 978-3-031-86715-6   ISBN 978-3-031-86716-3 (eBook)
https://doi.org/10.1007/978-3-031-86716-3

© The Editor(s) (if applicable) and The Author(s), under exclusive license to Springer Nature Switzerland AG 2025

This work is subject to copyright. All rights are solely and exclusively licensed by the Publisher, whether the whole or part of the material is concerned, specifically the rights of translation, reprinting, reuse of illustrations, recitation, broadcasting, reproduction on microfilms or in any other physical way, and transmission or information storage and retrieval, electronic adaptation, computer software, or by similar or dissimilar methodology now known or hereafter developed.
The use of general descriptive names, registered names, trademarks, service marks, etc. in this publication does not imply, even in the absence of a specific statement, that such names are exempt from the relevant protective laws and regulations and therefore free for general use.
The publisher, the authors and the editors are safe to assume that the advice and information in this book are believed to be true and accurate at the date of publication. Neither the publisher nor the authors or the editors give a warranty, expressed or implied, with respect to the material contained herein or for any errors or omissions that may have been made. The publisher remains neutral with regard to jurisdictional claims in published maps and institutional affiliations.

This Springer imprint is published by the registered company Springer Nature Switzerland AG
The registered company address is: Gewerbestrasse 11, 6330 Cham, Switzerland

If disposing of this product, please recycle the paper.

To all geoscience researchers who are interested in field geology and outcrop studies, palynology, structural geology and petroleum geology

Preface

In palynofacies origin, organic thermal maturation and petroleum migration pathway model synthesis exposé herein; the source rock potential and structural traps model triggered by the kingdom animalia and natural plant processes, paleoenvironment and kinematics systems of faults substantiates the quality of crude oil, petroleum migration pathway and hydrocarbon prospectivity of any sedimentary basin. This book highlights a detailed high-resolution palynofacies origin and depositional environment, source rock potential, hydrocarbon migration pathway and structural hydrocarbon entrapment mechanisms of the outcrop lithostratigraphic units of the Enugu Formation of the inland Anambra basin and the Danian units. The geographic, geomorphic and physiographic setting of the study plays a vital role in the examined accessibility of the outcrop locations of the study, embedded within the Enugu metropolis, Neke Isi-Uzo, Ugwogo-Nike area and the vicinity of Ikpankwu. The escarpment topographic dimensions with a visible undulating rolling topography along with several highlands and drainage systems physiographic attributes of the Enugu area escarpment terrane are enclaved within latitude - 6°00′ N to 7°00′ N and longitude 7°00′ E to 7°45′ E.

Hydrofluoric (HF) acid used for the removal of silicates and HCl that is instrumental for the removal of carbonates was important for the removal of carbonate in calcareous mudrock facies samples when necessary. The digested samples were sieve-washed, dispersed in polyvinyl alcohol and mounted on glass coverslips and slides before microscopic studies, and visual identification of particulate organic matter was undertaken. The visualized abundance and hydrodynamics quality of the palynofacies debris are substantiated by the abundance of relatively well-preserved yellow and dark brown amorphous organic matter, marine taxa, opaque particles, with diverse size ranges of dark brown structured phytoclasts and other terrestrial palynomorphs.

The Post Santonian origin of the Anambra Basin buttressed the tectonic events responsible for the evolution of the Anambra Basin and the stratigraphic framework of the basin substantiated by the Nkporo Group, Mamu, Ajali and Nsukka formations. Texturally mature coarse- to fine-grained sandstones, variable sedimentary structures, internal bed geometries, depositional contacts and nature of bedding are the prognostic sedimentary grain attributes of Aguabo pedestrian bridge ridge section, Neke Isi-Uzo road cut ridge, Amagu section, Ozalla/Four-Corner road Cut and the Ikpankwu quarry section.

The depositional environment processes of the Danian outcropping units and studied Enugu Formation sediments in this book authenticate the palynofacies origin and provenance, depositional environment and hydrocarbon production potential within the context of the basin. The quality and quantity of palynofacies particles and profusion of land-derived microflora over marine dinocysts illustrate the dynamic paleoenvironment processes of transitional marine settings of the inner neritic zone with intermittent outer neritic influences consistent with estuarine, nearshore and freshwater environments, whereas the Danian units demonstrate shallow marine settings with deep marine influenced macro environments, embedded as products of lower and upper deltaic plains.

These paleoenvironment indices are products of palynofacies elements idealized in thermally mature Type II-III kerogen indicative of oil and gas prone hydrocarbons for the formations. Hanging-wall closures, footwall closures, half graben, horst, collapsed-crest-graben of conjugate fault structural traps styles along with infilled and open joint structures within the faulted and unfaulted sedimentary rock sequences highlight the structural hydrocarbon entrapment mechanisms inherent in the Enugu Formation and Danian outcropping sediments.

Primary and secondary petroleum migration pathway styles in rock pores and fracture sequences are the key processes in hydrocarbon entrapment in southeastern Nigerian sediments. In palynofacies imprints and petroleum migration pathways model of this book: the source rock potential and structural traps model triggered by the paleoenvironment and kinematics systems of faults substantiates the quality of crude oil, petroleum migration pathway to visible reservoir rocks and hydrocarbon prospectivity of the earth's sedimentary basins.

This book stemmed from series of rigorous detailed outcrop studies and palynological research at the Department of Geology, University of Nigeria Nsukka. The target audience of this book includes palynologists, micro and macro paleontologists and biostratigraphers, archeologists, structural geologists, petroleum geoscientists, petroleum engineers and other geoscientists interested in understanding the basic concepts of hydrocarbon generation, maturation system and migration from the shale source rock or shale reservoirs to recovery from the basic reservoir rocks.

Nsukka, Nigeria   Kachikwulu Kingsley Okeke

Acknowledgements The author is most grateful to God in the communion of the Angels and Saints for safe keep from ghastly accident and series of hospital treatments. Special thanks to my mother Florence Odukwu Okeke-Anakputa, siblings—Ifedioramma, Nnenne, Chidiebele, Ebuka, Chukwuemeka, Odirachukwumma, Chinemelum and Igwe Onyebuchi Patrick Okeke-Anakputa family. Thanks to Profs. O. P. Umeji, A. W. Mode, S. C. Obiora, Mrs N. A. Ulasi and Dr. C. O. Maduewesi of Department of Geology, University of Nigeria, Nsukka, for immerse contributions that led to the publication of an aspect of this work. I am very grateful to my wife Ukamaka Debora Iroko for assistance during various stages of geologic field studies and for providing me with unfailing support and continuous encouragement.

Competing Interests The author has no competing interests to declare that are relevant to the content of this manuscript.

Contents

1 **Source Rock Potential, Palynofacies depositional Environment Synthesis and Structural Traps, of the Outcropping Lithostratigraphic Units of the Anambra Basin: Petroleum Migration Pathway and Hydrocarbon Prospectivity** 1
   1.1 Introduction 2
   1.2 An Overview of Anambra Basin Southeastern Nigeria 4
        1.2.1 Geographic, Geomorphic and Physiographic Setting 0
        1.2.2 Review of Literatures 0
   1.3 Aim and Objectives of Study 8
   1.4 Methodology 9
        1.4.1 Sample Collection 0
        1.4.2 Palynological Sampling Technique and Processing 0
        1.4.3 Logging 0
        1.4.4 Quantitative Statistical Analyses 0
   References 15

2 **Regional Geology and Basin Evolution** 19
   2.1 Geologic Framework 20
   2.2 Regional Tectonic Setting 21
   2.3 Regional Stratigraphy 26
   References 27

3 **Outcrop-Based Studies and Lithologic Description of Outcropping Stratigraphic Successions** 29
   3.1 Field Description 30
        3.1.1 Aguabo Pedestrian Bridge, Ridge Section (Outcrop 1) 31
        3.1.2 Neke Isi Uzo Road Cut Ridge (Outcrop 2) 32
        3.1.3 Amagu Section by Enugu-Port Harcourt Roadcut
               (Outcrop 3) 35
        3.1.4 Ozalla/Four-Corner Road Cut (Outcrop 4) 38

		3.1.5 Ikpankwu Ridge Section (Outcrop 5)	38
	3.2	Synopsis of the Lithologic Units Field Physiognomy and Facies Change Structural Trap Paradigm	39
4	**Palynofacies Depositional Environment, and Source Rock Potential of the Enugu Formation and the Danian Outcropping Lithostratigraphic Units**		41
	4.1	Origin and Palynofacies Constituents Types of the Enugu Formation and the Danian Lithostratigraphic Units	42
		4.1.1 Amorphous Organic Matter (AOM)	44
		4.1.2 Resin	45
		4.1.3 Structured (Wood) Phytoclasts	46
		4.1.4 Cuticle Phytoclast	47
		4.1.5 Unstructured (Degraded) Phytoclasts	47
		4.1.6 Opaque Debris	47
		4.1.7 Palynomorph Microfossils	50
	4.2	Metamorphosis and Organic Matter Degradation	50
	4.3	Paleoenvironment	51
		4.3.1 Palynofacies Depositional Environment of the Enugu Formation	52
		4.3.2 Palynofacies Depositional Environment of the Danian Lithostratigraphic Units	55
	4.4	Source Rock Potential	58
	References		63
5	**Tectonics and Structural Geology of Hydrocarbon Traps**		67
	5.1	Tectonics and Structural Traps Synopsis	67
	5.2	Fault Systems of the Enugu Formation, of the Anambra Basin and Danian Lithostratigraphic Units, Southeastern Nigeria	70
		5.2.1 Anticlinal Structures	71
		5.2.2 Growth Faults	76
		5.2.3 Footwalls	76
		5.2.4 Hanging Wall	77
		5.2.5 Stratigraphic Traps	78
	5.3	Fault Systems and Petroleum Migration Pathway of Enugu Formation of the Anambra Basin and Danian Units in Southeastern Nigeria	79
	5.4	Kinematics of Faults and Joints in the Enugu Formation of the Inland Anambra Basin and Danian Units of Southeastern Nigeria	83
	5.5	Clinoform Morphology and Structural System of the Anambra Basin	86
	References		88
Conclusions			91

Chapter 1
Source Rock Potential, Palynofacies depositional Environment Synthesis and Structural Traps, of the Outcropping Lithostratigraphic Units of the Anambra Basin: Petroleum Migration Pathway and Hydrocarbon Prospectivity

Abstract Detailed high-resolution palynofacies hydrodynamic and depositional environment, source rock potential and structural hydrocarbon entrapment mechanisms of the outcrop lithostratigraphic sections demonstrates the synergy between the formation, production and hydrocarbon accumulation processes. The analyzed outcroping lithostratigraphic units of the Enugu Formation of the Anambra Basin and Danian units revealed the structural trap style, hydrocarbon production potential framework and migration pathway for oil and gas reserves of the formation for hydrocarbon exploration and exploitation. The well-established lithostratigraphic framework of the Late Campanian to Danian Anambra Basin consists of the Enugu Formation of the Nkporo Group, Manu, Ajali and Nsukka formations. The Anambra Basin is an outcome of the Santonian tectonic event that affected the Benue Trough where the compressional uplift of sediments created a depocenter for the accumulation of the sedimentary sequences of the Anambra Basin formations. The sedimentary characteristic attributes of the Late Campanian to Early Maastrichtian Enugu Formation consist of the carbonaceous shale and sandstone whereas the Danian units is denoted by dark carbonaceous shale and sandstone units. The sedimentary rocks lithostratigraphic units of this rock cycles was illustrated as deposits of regressive and trangressive eustatic sea level fall and rise cycles deposited in fluviotidal, deltaic, shelfal and deep marine settings. The exuberance and hydrodynamics quality of the visual palynofacies analysis is substantiated by the abundance of relatively well-preserved yellow and dark brown amorphous organic matter, marine taxa, opaque particles, with diverse dark brown structured phytoclasts sizes and terrestrial palynomorph. The palynofacies components illustrates dynamic paleoenvironment of transitional marine settings of the inner neritic zone with intermittent outer neritic influences consistent with estuarine, nearshore and freshwater environments based on the normal varied quality and quantity of palynofacies constituent and profusion of land-derived microflora over marine dinoflagellate cysts for the Enugu Formation. The depositional array of the Danian units indicates shallow marine settings with deep marine influenced macro environments, embedded as products of lower and upper deltaic plains with similar variable palynofacies standard inherent in the Enugu Formation. Palynofacies elements suggest a thermally mature Type II–III kerogen indicative of oil and gas prone hydrocarbon for the formations. The kerogen types are conceptualized in the Enugu Formation's oil seepages and natural gas flare. The natural gas flares reflect the Kerogen Type III (gas prone) palynofacies constituents, while the oil seepages are linked with the Kerogen Type II palynofacies groups. The key structural traps of hanging-wall closures, footwall closures, half graben, horst, collapsed-crest-graben of conjugate fault array, along with infilled and open joint structures within faulted and unfaulted sandstone sequences exhibits the structural hydrocarbon entrapment mechanisms of the Enugu Formation and Danian outcropping sediments. Primary and secondary migration petroleum pathway in sedimentary rock pores and fracture sequences are the key primary means of creating petroleum accumulations in the Anambra Basin. The oil seepages and natural gas flares of the study reveal the structural style elements of the formations as a petroleum system design for hydrocarbon entrapment and preservation model within the Anambra Basin. The structural style complexity, lateral and vertical facies changes, and hydrocarbon prospectivity authenticates the oil seepages and gas flare prediction and subsequent petroleum crude oil and gas drilling pathway as well as exploration and exploitation campaign in the Anambra Basin.

Keywords Oil seepage · Hydrocarbon entrapment · Enugu Formation · Nsukka Formation · Depositional environment · Danian sediments · Anambra Basin · Primary migration and Secondary migration

1.1 Introduction

The acquisition of seismic logs in oil exploration campaign initiated the appraisal of seismic logs and subsurface stratigraphic data as the authentic concept in interpreting the sedimentary structural trapping style, mechanics, oil migration pathway, palaeoenvironment and lateral facies changes of sedimentary rocks. The frequently surpassed outcrop studies and interpretation of structural trap quality of sedimentary rocks in outcrop structural geology analysis during exploration and exploitation campaign stands out as the primary aspect of geology instinct in hydrocarbon exploration and reservoir models. Outcrop studies are significant in the development of the qualitative and quantitative descriptions of sediment packages and rock characteristics. However, seismic data and well logs provide outstanding structural and stratigraphic information on subsurface but relatively lack the potential to resolve small-scale vertical and lateral attributes of lithofacies. The hydrocarbon reserves of the Anambra Basin is ranked next to the globally acclaimed hydrocarbon rich Niger Delta Basin due to the quality and quantity of hydrocarbon exploration and exploitation in the basins. Anambra Basin is amongst the inland intracratonic basin located adjacent to the Niger Delta terrane of Nigeria.

Photosynthesis is the substantial medium for origin and growth of algae and other micro and macro plants which are the direct progenitors of organic matter and hydrocarbon resources. The abundance, occurrence and degree of maturation

1.1 Introduction

of particulate organic matter inherent in sedimentary rock depend on the paleoenvironment and depositional processes, predominant during the sediment deposition and diagenetic processes. The quantity and quality of palynofacies elements are useful in the interpretation of palynofacies hydrodynamics and depositional environment. This is because classification and quality of the individual palynofacies constituents current in clastic rocks is key to effectual kerogen analysis of such rocks. The palynofacies model of the Anambra Basin improved the understanding of particulate organic matter occurrence which can be linked to the paleoenvironment and palynofacies hydrodynamic interpretation, development of oil fields and hydrocarbon prospectivity of the basin for conventional and unconventional oil and gas exploration (Table 1.1).

Petroleum geology is a profoundly synthesised geological discipline that is of greatest significance for finding and recovering oil and gas as detailed in Chaps. 1 to 5 of this book. In the first chapter, palynofacies origin, organic thermal maturation and petroleum migration pathway model were reviwed and documented. The appraisal of the palaeontological and natural plant data processes, paleoenvironment, kinematics systems of faults, palynological acid maceration technique and the geographic, geomorphic and physiographic setting characteristics of the study are also presented in Chap. 1. The second chapter presents the regional geology, basin forming tectonics and basin fill expatiated in the regional stratigraphic setting. The field geologic description of outcrop sections and the origin and palynofacies constituents types along with palynofacies depositional environment and source rock potential of the study were amongst the main focus of this book detailed in Chaps. 3 and 4 respectively. In departure from the source rock potential highlighted in Chap. 4, tectonics and structural geology of sedimentary structural trap hold forth about the fault systems, fault systems and petroleum migration pathway along with kinematics of faults and joints of this book are presented in Chap. 5. Chapter 5 ends with a discussion on the clinoform morphology and structural system of the Anambra Basin.

Table 1.1 Some terminology of palynofacies sample and geology analysis

Terminology	Definition
Geomorphology	This is the scientific study of the evolution, origin, and form and processes of distribution of landforms of the earth's crust
Acid maceration	The palynological acid maceration technique of using hydrochloric (HCl) acid for removal of carbonates and hydrofluoric (HF) acids to remove silicates during sedimentary rock digestion
Physiography	Synonymous with physical geology, which deals with the study of physical features of the earth's surface
Kinematic examination	The interpretation and demonstration of tectonic motion and direction of rocks during undeformed to deformed transformation state

1.2 An Overview of Anambra Basin Southeastern Nigeria

1.2.1 Geographic, Geomorphic and Physiographic Setting

Geographically, the study area is part of the Anambra Basin location within the Enugu metropolis, Neke Isi-Uzo, Ugwogo-Nike area and Ikpankwu. It is bounded by latitude 6°00′ N to 7°00′ N and longitude 7°00′ E to 7°45′ E, which covers an outcrop area of approximately 828.75 km^2.

The study area and outcrops are accessible through footpaths, seasonal paved and unpaved main roads connecting Nsukka with Enugu vicinity down to Okigwe and Umuasua domain traversing Enugu, Imo and Abia States of southeastern Nigeria. The outcrop localities within this area include Umuasua, Ikpankwu, Uturu, Ihube, Inyi, and Oji, Enugu, Neke Isi-Uzo-Nsukka, Nike in the vicinity of Nsukka Formation and Enugu, Nkporo, Leru, Okigwe, areas of the Enugu and Nsukka formations (Fig. 1.1).

Geomorphologically, the study area exhibits an escarpment topographic dimensions with an undulating rolling topography that is encountered in the vicinity of Enugu from Enugu-Ezike down to Agbani and Okigwe with a long steep slope. The topography view is controlled by the lithologies and underlying bedrock morphotypes of the Anambra Basin pronounced in the long steep slope. The topographic elevation maximum peaks of more than 480 m was ascertained from the 3D Digital Elevation Map (DEM) within the sandstone ridges of the Enugu domain which form a systematic topographic high from Enugu Ezike to Nsukka and Opi through Ukehe, Enugu and Agbani down to the Uturu and Okigwe vicinity. These highlands are the physiographic attributes of the topography (relief) and drainage systems of the Enugu area escarpment terrane.

These are the topographic highs of the Enugu escarpment ridge formed by sandstones ridges with visible gradual ascent from western and eastern low-lying plains of the mapped area to series of the escarpment highland of the sandstone ridge in the northern and southern parts and direction of the study area. The low-lying plains of the mapped area are characterized by shale and other mudrock sandstone lithologies. The drainage fluvial system of the Ekulu, Ogbete, Asata, and Iva Valley rivers are the primary streams with typical dendritic drainage pattern attributed to a massive lithological uniformity and erosion resistance. The 3Dimensional Digital Elevation Map (DEM) punctuates the major escarpment ridge systems (Fig. 1.2) of the study area.

1.2.2 Review of Literatures

The discovery of coal triggered significant geological and exploration curiosity in the Anambra Basin since the early 1900 to Recent. The recent natural gas flare (Okeke et al. 2023a, b, c) and oil seepages (Nwajide 2013, 2022; Nwajide and Reijiers 1996) in the Enugu Formation of the Nkporo Group triggered the scientific analysis of the source rock potential, sedimentary structures trapping mechanism and

1.2 An Overview of Anambra Basin Southeastern Nigeria

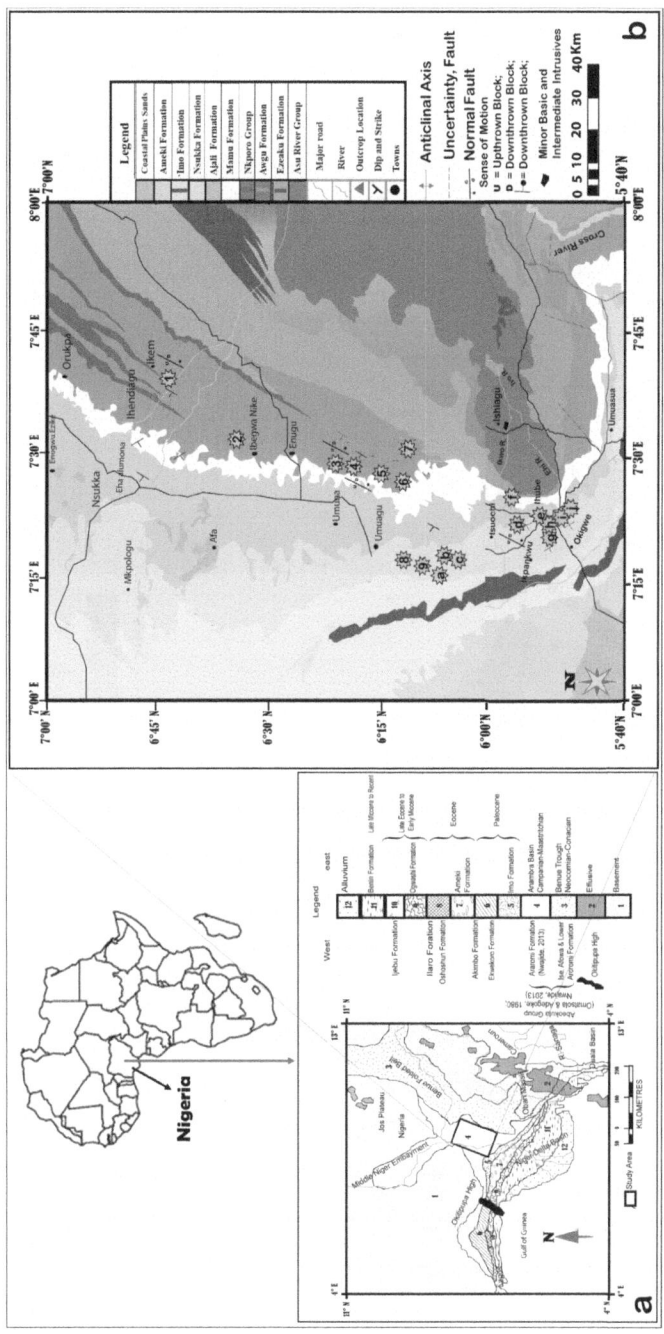

Fig. 1.1 Generalized geological map of the study area showing outcrop locations, accessibility, dip and strike and the faulted sections of the studied outcropping Danian sections and the Enugu Formation of the Anambra Basin, southeastern Nigeria

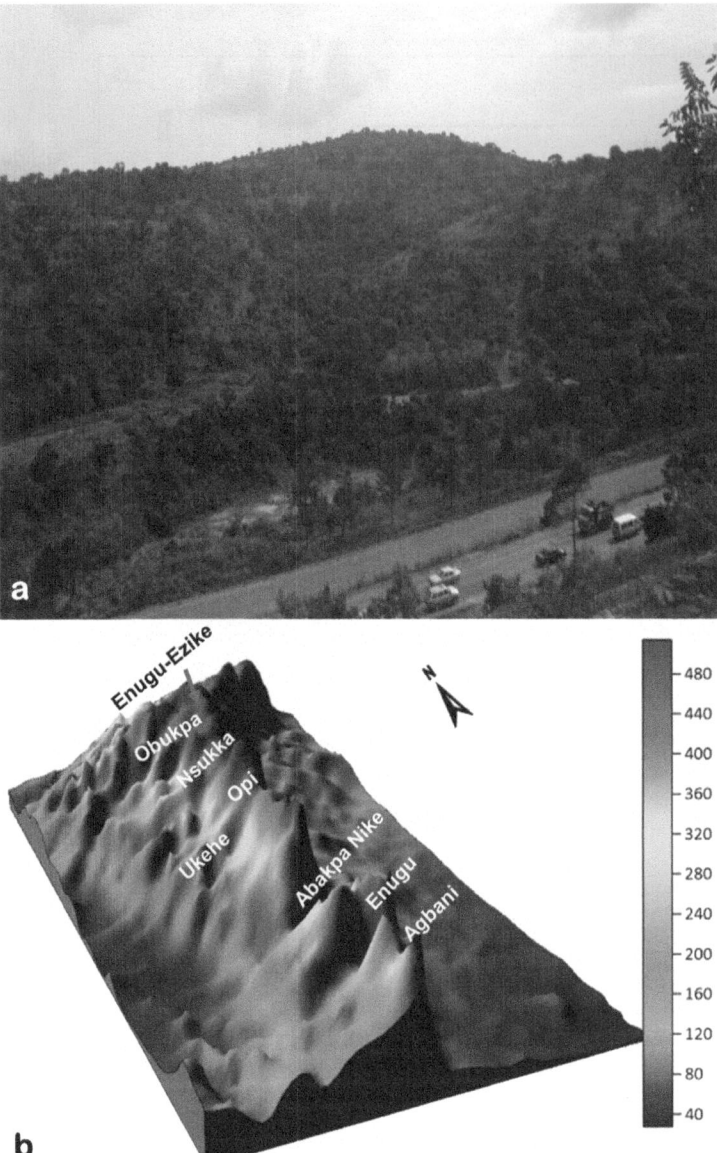

Fig. 1.2 a Topographic view of the of the scarp face of the Enugu cuesta illustrating the long steep slope, escarpment highland and low lying plains of the Enugu cuesta of southeastern Region, Nigeria. **b** 3D Digital Elevation Map (DEM) of the study area showing the escarpment features of the sandstone ridge with the highest 480 m elevation seen at the northern and southern part of the study area (note: colour code bar is in meters)

1.2 An Overview of Anambra Basin Southeastern Nigeria

petroleum migration pathway of oil and gas hydrocarbon system of the Enugu Formation and Danian section of the Nsukka Formation. A concise geologic framework of the tectonic evolution of the Anambra Basin, southeastern Nigeria was reported by Nwajide (2013); Umeji (2000); Murat (1970, 1972); Benkhelil (1986); Ekweozor (1982); Hoque and Nwajide (1985), and Ojoh (1992). The Anambra Basin was formed from the second tectonic compressional uplift phase of the Abakaliki-Benue Trough during the Santonian. The displacement of the depositional centre from the Abakaliki Basin to the down warped Anambra Basin engineered the evolution of the sedimentary rock formation from Late Campanian to Danian. The deposited sedimentary rock units includes the Nkporo Group, Mamu, Ajali and Nsukka formations of Late Campanian–Danian times. These formations within the Anambra Basin projects sedimentary rock characteristics of carbonaceous shales and sandstone of the Enugu Formation of the Nkporo Group; alternating sandstones, sandy shales and interbedded sub-bituminous coal seams of the Mamu Formation. The Ajali Formation is characterized by fine- to coarse-grained sandstone with interbeds of clay laminae whereas the dark carbonaceous shales and sandstones with series of thin coal seams are products of the Nsukka Formation. The paleoenvironment reconstruction of the Anambra Basin demonstrates imprints of eustatic sea level rise and fall cycles deposition whereas the Enugu Formation was deposited in inner neritic zone with intermittent outer neritic influences. Mamu Formation was deposited in alternating continental to coastal swamp and inner neritic environment (Okeke 2023, 2024; Nwajide 2013, 2022; Dim et al. 2017; Umeji 2000) while the Ajali and Nsukka formations were deposited in fluvio-tidal, deltaic, shelfal and alternating coastal and inner neritic settings along with deep marine influences respectively (Okeke 2024; Nwajide 2013, 2022; Dim et al. 2017, 2020; Umeji 2000).

The kinematics of the faults and joints system of the Enugu Formation in the vicinity of Enugu were reported by Okeke et al. (2023a, b, c), Amogu et al. (2011) and Anigbogu (2013). The hanging wall closures, footwall closures, horst and collapsed-crest structure structural trap style of the outcropping and seismic imprints of the Enugu and Nsukka formations were documented (Okeke et al. 2023a, b; Dim et al. 2020, 2017; Amogu et al. 2011; Anigbogu 2013). However, a similar structural architecture of sedimentary basins was reported in the oil prone Niger Delta Basin and other areas of the earth through seismic data to understand the structural dynamics of footwall and hanging wall closures, anticlinal dip closures, faulted rollover anticlines, collapsed-crest structures, sub-detachment structures and horst block structural imprints (Doust et al. 1990; Doust and Omatsola 1990; Dim 2017; Dim and Onuoha 2017; Nwajide 2013; Orife and Avbovbo. 1981; Okeke et al. 2023a, b). The mechanism and pathway of primary and secondary petroleum migration of some formations in the Anambra Basin were first documented by Okeke et al. (2023a, b) to discern the pore scale distribution, expulsion and non-expulsion of hydrocarbons within open or closed pore spaces respectively. The systematic sole and integrate sedimentology, petrophysics, organic geochemistry, and statistical methods were stipulated and utilized to authenticate scientific and petroleum geology processes of primary and secondary petroleum migration models of the Anambra Basin (Okeke et al. 2023a, b, c) and other areas of the World (e.g., Mann 1994; Durand 1983, 1988; Ozkaya 1988; England et al. 1987; Ungerer 1990; Price 1989; Welte 1987).

Many detailed studies on the biostratigraphy and depositional settings of the Enugu and Nsukka formations outcrop sections with the aid of micropaleontology tools of foraminifera, palynology and other macropaleontology evidence of gastropod, bivalve, ammonite, and fish teeth can be found in the works of Zarboski (1982, 1983), Umeji (2000, 2006, 2011), Agagu et al. (1985), Rayment (1965), Nwajide (2013, 2022), Okeke et al. (2023a, b, c), Kelechi et al. (2023), Chiaghanam et al. (2012) and Chiadikobi et al. (2018).

Palynology works of the Enugu Formation (Okeke 2023; Umeji 2006, 2011, 2000) and its lateral equivalents in the Calabar Flank (Edet and Nyong 1993, 1994; Salami 1990) and southwestern Nigeria (Oloto 1990) were documented to harness the stratigraphic position and sedimentary environment of deposition. These studies pinpointed Late Campanian-Early Maastrichtian as the age of the Enugu Formation of the Anambra Basin.

Few scholarly advances have been made in the understanding of palynofacies elements, kerogen maturation, palynofacies depositional environment reconstruction and hydrodynamic interpretation (Umeji 2006, 2011, 2000; Okeke 2024; Okeke et al. 2023a, b). Several palynofacies schemes and description of the origin and characteristics of palynofacies elements of Nigerian sedimentary basins were previously itemized (Okeke 2024; Okeke et al. 2023a, Okeke and Umeji 2018a, b). The palynofacies thermal alteration of opaque palynodebris and other particulate organic matter elements are products of oxidation of translucent woody angiospermous and gymnospermous plant remains during prolonged hydrodynamic transport and post-depositional alteration of terrestrial plants inherent in other terrestrial phytoclast groups (Okeke and Umeji 2018a, b).

The palynofacies zones and unconformity exposé indicated that the Enugu Formation of the Nkporo Group was deposited in a transitional marine setting consistent with estuarine, nearshore and freshwater environment with terrestrially derived palynodebris (Umeji 2006, 2011; Okeke et al. 2023a, b, c; Kelechi et al. 2023). The paleoenvironment reconstruction of the Danian chronostratigraphic section of the Nsukka Formation posits shallow marine settings with deep marine influences (Umeji 2000, 2005; Umeji and Nwajide 2007; Umeji and Edet 2008; Okeke 2024).

1.3 Aim and Objectives of Study

This research aims to understand the depositional environment, source rock potential, sedimentary structure trapping orientation and configurations, structural mechanics and lateral facies changes of the studied Enugu and Nsukka formations of the Anambra Basin. In this sense, this work focuses on the appraisal of the sedimentary structural trapping, hydrocarbon production potential and petroleum migration pathway of sedimentary rocks.

The principal objectives of this research are to:

- Interpret the depositional environment and lateral facies changes based on palynofacies hydrodynamics.
- Produce a detailed kerogen type and thermal maturation model for potential conventional and unconventional hydrocarbon prospects and contribute more information to oil exploration campaigns in the basin.
- Highlight the key structural trap style to elucidate the link between the source rock and sedimentary structures trapping mechanism of the petroleum system
- Delineate the migrating pathway of the Anambra Basin's hydrocarbon natural gas flare and oil seepages of the Anambra Basin.
- Illustrate the kinematics systems of faults and joints in the studied formations

1.4 Methodology

Sedimentary rock outcrops with suitable structural configurations were examined based on the synthesised workflow plan exhibited in this work. Proper measurement of the trend of faults and joints with Brunton Compass and a Global Positioning System (GPS) aided the outcrop scale structural analysis. The Garmin GPS also engineered the recording of geographic coordinates of the outcrop locations on the map shown in Fig. 1.1. The Anambra Basin's Minna geodetic datum georeferenced base map was activated using MapInfo Geographical Information System (GIS) platform which was instrumental in the systematic coordinates input.

The structural characteristic of the rocks were utilized for trap classifications, whereas its effects on hydrocarbon exploration and exploitation were discussed in this study. Sedimentary structures and structural styles are fundamental elements of the petroleum system inherent in sedimentary and igneous rocks entrapment mechanisms. Outcrops from stream sections, gutter excavations, gully sites, road cuts and quarry sections exposed well preserved rock units of the Enugu and Nsukka formations across Nsukka through Ugwogo-Nike, Enugu plains, Ikpankwu sandstone and Okigwe ridges of the Anambra Basin.

The geology field mapping provided good geological sections for logging, classification and sampling of outcropping sediments for palynofacies analysis of various lithostratigraphic units. The attitude of bed (strike, dip direction and dip amount), rock types, sedimentary facies types and other sedimentary rock information of thickness, rock geometries, sedimentary structures, nature of bedding and texture of the rock types guided the production of a detailed geologic map of the study area. The study of the collected rock samples was subjected to sedimentary rock description and palynology acid maceration technique for palynofacies analysis. The key descriptive sedimentary rock attributes include lithology (rock type), texture (grain size, grain shape and grain sorting), micro and macro fossil content, ichnofossils and sedimentary structure.

Palynofacies laboratory analysis utilized in this work commenced through detailed sedimentary geology field outcrop studies of the Enugu Formation of the Nkporo Group and the Danian section of the Nsukka Formation. This was supported by

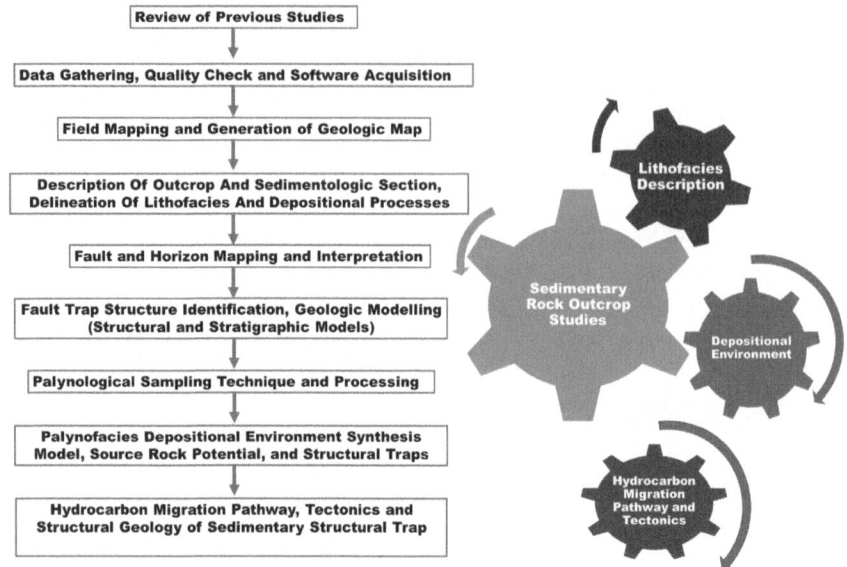

Fig. 1.3 Generalized workflow for this research work

detailed desktop descriptions of depositional sequences, sedimentary structures and plotting of grain-size data (Fig. 1.3). The outcrop field interpretations have been supported by the collection of well-preserved siltstone, mudstone and shale samples for kerogen analysis in the Palynological Biostratigraphy Labouratory, Department of Geology, University of Nigeria, Nsukka. The systematic techniques utilised in this research are (i) sample collection, (ii) palynological sampling and processing, (iii) logging, and (iv) quantitative statistical analyses.

1.4.1 Sample Collection

The effects of oxidation and particle size of palynomorphs necessitated that precise samples were collected from fresh outcrops on road cut, gully, stream, quarry and river sections from bottom to top to avoid contamination by extant and older palynomorphs. Where fresh samples were unavailable digging was done to expose them to good, fresh surfaces. The well-preserved samples were subjected to standard acid palynological sample processing using Hydrochloric acid (HCl) and Hydrofluoric acid (HF) acids.

1.4.2 Palynological Sampling Technique and Processing

Palynological Processing

The samples for this work were prepared at the Palaeontology and Biostratigraphy laboratory, Department of Geology, University of Nigeria, Nsukka using the standard palynological processing procedures. Each laboratory worldwide develops its own processing methods concerning cost. The method used here is a hybrid of palynological laboratory processing methods proposed by Traverse (1988), Jansonius and Mcgregor (1996), Moore and Webb (1978), Okeke (2017), Enuwnwemba (2018), Umeji (2007) and Ikegwuonu (2013). The outcrop samples prepared were shale, limestone, siltstone, sandstone in the absence of lignite and coals. The sample preparation involved three main phases/stages: mechanical, chemical and residue mounting.

1.4.2.1 Mechanical Stage

Cleaning and Removal of Field Contaminants

The samples were scraped clean with a penknife to remove any contaminations or encrustations of fungus and algae due to aerobic conditions that encourage microbial activities. The encrustations occur in yellow, white and green staining which is characteristic of fungus spore of different botanical origins.

Crushing of Samples

The samples were carefully cleaned before 10 g were obtained from shale, mudstone, limestone and sandstone samples for processing. The accurate weighing of 10 g of the sampled sediments is to avoid maceration of excess and unnecessary sediments. They were crushed to a surface area of 1–2 mm size fraction using agate mortar and pestle. Crushing increases, the surface area of the rock samples for chemical action and prompts maceration of the sediments. Finer crushing was avoided because it breaks the larger palynomorphs and palynofacies elements while coarser crushing leaves the core of the grains not acted upon by acids.

The mortar and pestle were washed with water, detergent, sponge and brush after each crushing to prevent contaminations as many samples were crushed successively with the same mortar and pestle. The mortar and pestle were dried before another round of crushing to prevent the samples from getting wet and sticking to the surface of the mortar and pestles due to the plasticity of the shale and mudstone samples.

1.4.2.2 Chemical Stage

Hydrochloric Acid (HCl) Treatment—Removal of Carbonates

Hydrochloric acid (HCl) treatment is to remove carbonates. 50 ml of 35% HCl was added from the acid dispenser, a few drops at first to test for $CaCO_3$ since the presence of calcium carbonate ($CaCO_3$) in the sediment can be detected by adding dilute HCl. If the observed reaction is violent, it indicates the presence of $CaCO_3$, If not proceed to the next stage. The reaction was allowed to subside before more of the acid was added. The rest of the acid was added and allowed to stand for 12–18 h. These precautions prevented the sample from foaming and spilling over.

HCl treatment dissolved the carbonate minerals including calcareous fossils leading to the concentration of palynofacies constituents. On completion of the reaction, the acidic supernatant was decanted. More water was added to the sample to dilute it until it became neutral. The decantation was done at two hourly intervals, until the supernatant solution turns the litmus paper neutral. Precaution was taken at the last decantation to avoid the loss of the sample, yet with very little water remaining in order to avoid diluting the next acid-Hydrofluoric acid (HF) to be added.

Hydrofluoric Acid Digestion—Removal of Silicates

Hydrofluoric acid (HF) digestion was done to remove the silicates. 50 ml of 40–48% HF was added to each carbonate free sample and left for 72 hours. It was stirred once a day to enhance digestion. The longer the duration, the more complete the digestion since this does not damage the palynofacies components. The dissolution of silicates was complete when all grittiness was lost. On completion of digestion, the HF acid was diluted with water before sieve washing disposal technique to avoid corroding the sink. The remaining sample was transferred into one (1) litre plastic beaker. Water was added to the residue in a beaker up to one litre level or more to dilute the HF to a safer concentration for washing. The samples were washed through 10 μ nylon sieve mesh under running tap water using a directed jet of water. The higher the pressure and intensity of the water running through the tap, the easier and less time it takes to wash off the acid and mud particles. The dissolved silicates, fine debris, acid, water and other substances broken down to less than 10 μ size were washed out until the litmus paper turned neutral. The nylon sieves were reused many times. To prevent contamination of the samples, the sieve was washed thoroughly after each use with water, detergent and plastic soft brush since hard brush will damage the surface of the sieve. An alkaline solution does not damage the sieve but acidic solution attacks the nylon.

All the samples were pipetted into 5 ml plastic vials, and kept for kerogen studies while all the samples digested for palynofacies analysis underwent these processes.

Kerogen

Kerogen studies were carried out on unoxidized organic matter. The unoxidized residue stored after HF digestion was used for the kerogen studies. In case of any slide damage or need for further study of the samples, one falls back to the kerogen sample which is equivalent to 10 g of the analyzed sample.

Preparation of Dispersant (Polyvinyl Alcohol)

Twenty grams of polyvinyl alcohol were dissolved by continuous stirring in 100 cl of warm water in a beaker. The stirring was continuous until the granules were completely dissolved, producing a gummy fluid (Umeji 2007; Okeke 2017). About

1.4 Methodology

5–8 drops of this fluid in 5 ml equivalent of the organic suspension was enough to keep the organic particles dispersed. The kerogen samples required 2–5 drops to be dispersed.

Dispersal of Organic Matter

Polyvinyl alcohol was used as the dispersant. Dispersal of organic matter was important because organic residue tend to cluster together unless suspended in a dispersant.

Mounting

After 1–2 days (24–48 hours) 3 ml water of the 5 ml equivalence of the 10 g aliquot labelled vial was pipetted out. The remaining 2 ml residue from the 10 g sample (siltstone, mudstone, shale and sandstone) was divided into 1 ml in each vial bottle (Fig. 1.4). 5 to 8 drops of polyvinyl alcohol dispersant were added in one of the 1 ml suspensions with a disposable plastic pipette and mixed thoroughly by shaking it. Check if the particulate organic matter elements are well dispersed; if yes stop and if no, add more drops until they are dispersed. Dense samples were diluted with more water to avoid clustering of palynofacies debris on the slide. One cover slips was placed on a neat flat surface. There are two types of cover slips, large (22 mm × 35 mm) and small (22 mm × 22 mm). A large cover slip (22 mm × 35 mm) was used.

At this juncture, some very well-dispersed samples were pipetted on the cover slip so that it covered the slip very well. Two coverslips of non-quantitative kerogen were also made from the stored non-oxidized aliquot. The slips were covered to avoid contamination with extant palynofacies constituents and left to dry at room temperature for some days. It took at least 1 to 2 days since the samples were prepared during the rainy season. It usually takes fewer days during the dry season. When dried, the slide for kerogen labelled K was placed on a neat surface. The slide was labelled from the left-hand side. A mixture of Araldite in equal proportions was used as a mounting medium. The mixture was spread on the centre of the slides with the tip of the rod such that it covered the 22 mm × 35 mm slip size range. The cover slip carrying the dried sample was gently lowered onto the slide starting from the far edge of the mountant. This was to avoid trapping of air bubbles beneath the cover slip. It was set such that the slide containing the palynofacies debris sticks to the araldite surface. The slide was lowered unto an electric stove which was used as a hot plate for some minutes depending on the temperature. The heat allowed the mounting medium and slip to set permanently on the slide. Precaution was taken to avoid burning the particulate organic matter and render the slide useless. It was observed that high temperatures could introduce air bubbles onto the slide which makes it very difficult to identify the palynofacies constituents on the microscope.

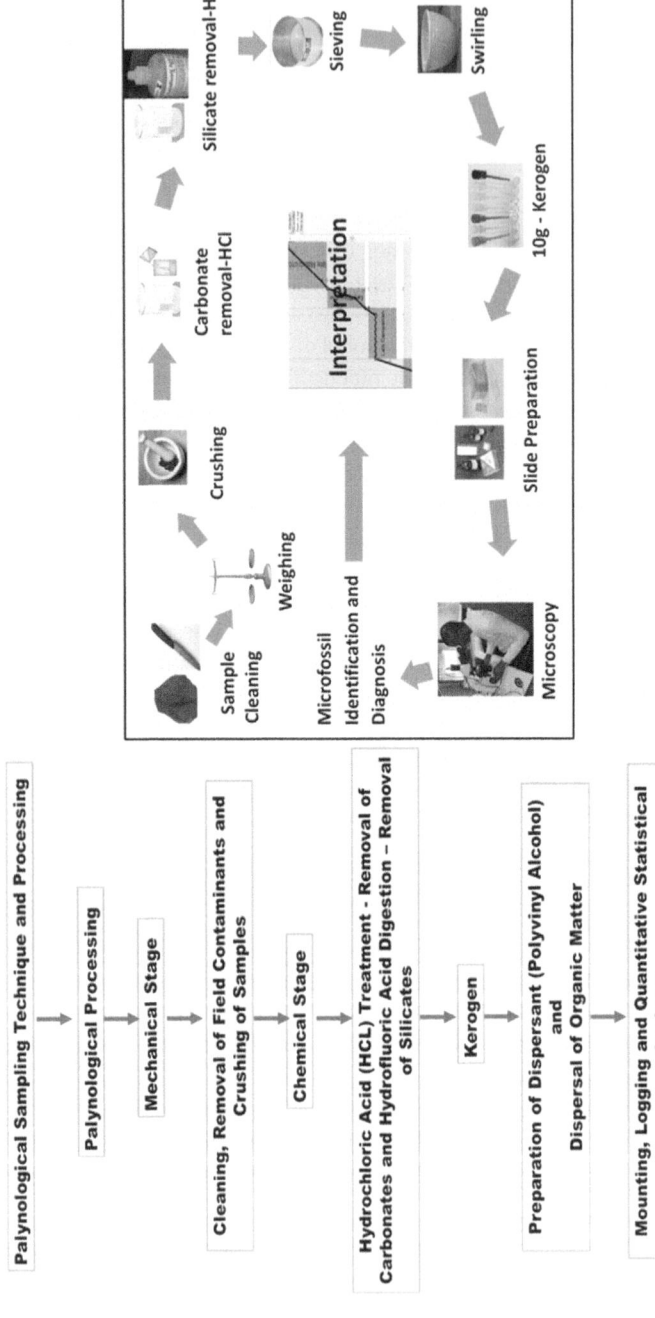

Fig. 1.4 Palynological processing workflow for this research work revealing a simplified palynological processing technique adopted in the work

1.4.3 Logging

Logging was done with the aid of a Motic B300 microscope. A minimum of five hundred grains were counted in each sample. The grains were taken from all scales of the slides because after the standard five hundred palynofacies count, another scale of the slides was scanned for new and rare forms of palynofacies particles that have not been counted. The slides were placed with the labelled part on the left-hand side of the microscope during the microscopic study. The absolute counts were converted to percentages to eliminate the differences in the counts.

1.4.4 Quantitative Statistical Analyses

A minimum particulate organic matter count of 500 kerogen particles (Figs. 4.3 and 4.4) was undertaken for every analysed sample. StrataBugs software (Strata-Data Ltd) was instrumental in the statistical data analysis where relative abundances and percentage charts are important for the systematic palynofacies geology standard interpretation. The following categories are frequently used in the text: few (1–5%), rare (5–10%), common (10–20%), abundant (20–40%) and superabundant (> 40%) (cf. Okeke 2023, 2024; Okeke 2017; Umeji 2006). The paleoenvironmental parameters of the study were obtained from the quantitative numerical examination of marine kerogen elements of dinocysts, foraminifera test linings and amorphous organic matter along with terrestrial pollen and spores microflora and other palynofacies elements of terrestrial origin calculated in percentages (%) of the total count and normal numerical count statistical measures.

The palynological Marine Index (PMI) of Helenes et al. (1998); calculated percentage (%) and normal count population/diversity distribution were integrated with the quantitative and qualitative particulate organic matter analyses of the formations in the Anambra Basin. These were used to express the relative abundances and diversity of palynofacies particles. Helenes et al. (1998) initiated Palynological Marine Index (PMI) statistical model to support the interpretation of depositional environments. PMI is calculated using the formula: $PMI = (Rm/Rt + 1)100$ where Rm is the richness of marine palynomorphs (dinoflagelletes, acritarchs and foraminiferal test linings), counted as the number of taxa per sample and Rt is the richness of terrestrial palynomorphs (pollen and spores) counted as the number of taxa per sample.

References

Agagu OK, Fayose EA, Petters SW (1985) Stratigraphy and sedimentation in the Senonian Anambra Basin of Eastern Nigeria. J Min Geol 22:25–36

Amogu DK, Filbrandt J, Ladipo KO, Anowai C, Onuoha K (2011) Seismic interpretation, structural analysis, and fractal study of the greater Ughelli Depobelt, Niger Delta basin, Nigeria. Lead Edge 30(6):640–648

Anigbogu CE (2013) Reservoir quality assessment of the Mamu formation in Enugu and its environs, southeastern Nigeria. MSc. thesis. University of Nigeria Nsukka, pp 1–200

Benkhelil J (1986) Caracteristiques structurales et evolution geodynamique du basin intracontinentale de la Benoue (Nigeria). Thesis d'etat, Nice, 275 pp

Chiadikobi KC, Chaghanam OI, Omyemesili OC, Omoboriowo AO (2018) Palynological study of the Campano-Maastrichtian Nkporo group of Anambra Basin, Southeastern, Nigeria. World News Nat Sci 20

Chiaghanam OI, Ikegwuonu ON, Chiadikobi KC, Nwozor KK, Ofoma AE, Omoboriowo AO (2012) Sequence stratigraphic and palynological analysis of the late Campanian to Maastrichtian sediments in the Upper-Cretaceous, Anambra Basin. A case Herngreen GFW, Chlonova AF. 1981. Cretaceous Microfloral Provinces. Pollen Spores 23:441–556

Dim CIP (2017) Hydrocarbon prospectivity in the eastern coastal swamp depo-belt of the Niger Delta Basin—stratigraphic framework and structural styles. Springer Briefs in Earth Sciences, Springer International Publishing AG, Gewerbestrasse, Switzerland, p 71

Dim CIP, Onuoha KM (2017) Insight into sequence stratigraphic and structural framework of the onshore Niger Delta Basin: integrating well logs, biostratigraphy, and 3D seismic data. Arab J Geosci 10(14):1–20

Dim CIP, Onuoha KM, Okeugo CG, Ozumba BM (2017) Petroleum system elements within the Late Cretaceous and Early Paleogene sediments of Nigeria's inland basins: an integrated sequence stratigraphic approach. J Afr Earth Sci 130:76–86

Dim CIP, Okonkwo IA, Anyiam OA, Okeugo CG, Maduewesi CO, Okeke, KK, Umeadi IM (2020) Structural, stratigraphic and combination traps on outcropping lithostratigraphic units of the Anambra Basin, Southeast Nigeria. Petr Coal 62(1)

Doust H, Omatsola E (1990) Niger Delta. In: Edwards JD, Santogrossi PA (eds) Divergent/passive margin basins, vol 48. American Association of Petroleum Geologists Memoir, pp 239–248

Doust H, Omatsola E (1990) Niger Delta. In: Edwards JD, Santogrossi PA (eds) Divergent/passive margin basins, vol 48. American Association of Petroleum Geologists Memoir, pp 201–238

Durand B (1988) Understanding of HC migration in sedimentary basins (present state of knowledge). Org Geochem 13:445–459

Durand B (1983) Present trends in organic geochemistry in research on migration of hydrocarbons. In: Bjoroy M et al (eds) Advances in organic geochemistry. Wiley, New York, pp 117–128

Edet JJ, Nyong EE (1993) Depositional environments, sea-level history and palaeobiogeography of the late Campanian-Maastrichtian on the Calabar Flank, SE Nigeria. Palaeogeogr Palaeoclimatol Palaeoecol 102(1–2):161–175

Edet JJ, Nyong EE (1994) Palynostratigraphy of the Nporo Shale exposures (Late Campanian-Maastrichtian) on the Calabar Flank, S. E. Nigeria. Rev Paleobot Palynol 80:131–147

Ekweozor CM (1982) Petroleum geochemistry application to petroleum exporation in Nigeria's lower Benue Trough. J Min Geol 19:122–131

England WA, Mackenzie AS, Mann DM, Quigley TM (1987) The movement and entrapment of petroleum fluids in the subsurface. J Geol Soc Lond 144:327–347

Enuenwemba LO (2018) Stratigraphic palynology and ammonite biostratigraphy of the late Maastrichtian-Selandian transition around Okigwe southeastern Nigeria, pp 1–172

Helenes J, De Guerra C, Vasquez J (1998). Palynology and chronostratigraphy of the upper cretaceous in the subsurface of the Barinas area, western Venezuela. AAPG bulletin, 82(7):757–772

Ikegwuonu ON (2013) Late Paleocene to early Miocene palynostratigraphy of sediments in Bende-Umuahia Area, Niger Delta Basin, Southeastern Nigeria, pp 1–168

Jansonius J, Mcgregor DC (1996) Introduction. In: Jansonius J, Mcgregor DC (eds) Palynology: principles and application, vol 1. American Association of Stratigraphic Palynologists Foundation, pp 1–10

Kelechi FM, Okeke KK, Ulasi N (2023) August. Geology and paleo-depositional environment of Leru and its environs, southern-eastern, Nigeria. In: SEG international exposition and annual meeting. SEG, pp. SEG-2023

Mann U (1994) An integrated approach to the study of primary petroleum migration. In: Parnell J (ed) Geofluids: origin, migration and evolution of fluids in sedimentary basins. Geological Society Special Publication No. 78, pp 233–260

Moore PD, Webb JA (1978) An illustrated guide to pollen analysis. Hodder and Stoughton, London, p 192

Murat RC (1970) Stratigraphy and palcogcography of the Cretaceous and lower tertian in southern Nigeria. In: Dessauvagie TFJ, Whiteman AJ (eds) African geology. University of Ibadan Press, Ibadan, pp 251–268

Murat RC (1972) Stratigraphy and paleogeography of the Cretaceous and lower tertiary in southern Nigeria. In: Dessauvagie TFJ, Whiteman AJ (eds) African geology. University of Ibadan Press, Ibadan, pp 251–266

Nwajide CS, Reijiers TJA (1996) Sequence architecture of the Campanian Nkporo and the Eocene Nanka Formation of the Anambra Basin, Nigeria. NAPE Bull 12(1):75–87

Nwajide CS (2013) Geology of Nigeria's sedimentary basins. CSS Bookshops Ltd., pp 347–411

Nwajide CS (2022) Geology of Nigeria's sedimentary basins. 2nd ed. Albishara Educational Publications, pp 394–443

Nwajide CS Hoque M (1985) Problems of classification and maturity evaluation of a diagenetically altered fluvial sandstone. Geologie en mijnbouw 64(1):69–77

Ojoh KA (1992) The southern part of the benue trough (nigeria) cretaceous stratigraphy, basin analysis, paleo-oceanography and geodynamic evolution in the equatorial domain of the south atlantic. Nape bull 7(2):131–152

Okeke KK, Umeji OP (2018a) Palynofacies, organic thermal maturation and source rock evaluation of Nanka and Ogwashi formations in updip Niger Delta Basin, Southeastern Nigeria. J Geol Soc India 92:215–226

Okeke KK, Umeji OP (2018b) Oil shale prospects of Imo Formation Niger Delta Basin, southeastern Nigeria: palynofacies, organic thermal maturation and source rock perspective. J Geol Soc India 92(4):498–506

Okeke KK, Mode A, Anigbogu EC, Umeadi IM, Odu NJ, Maduewesi CO, Ulasi NA (2023a) Source rock potential, palynofacies depositional environment synthesis and structural traps of the Enugu Formation Southeastern Region, Nigeria. Arab J Geosci 16(5):1–17

Okeke KK, Umeji OP, Dim CP, Ekwenye OC, Ulasi NA, Uwakwe OC, Maduewesi CO (2023c) Depositional facies and palynofacies provenance of clastic deposits: insight from Paleocene Strata in Southeast Region, Nigeria. Ir J Sci 47(1):73–90

Okeke KK, Mode AW, Eradiri JN, Umeadi IM, Maduewesi CO, Ulasi NA (2023b) Palynofacies depositional environment, source rock potential and structural traps on outcropping lithostratigraphic units of the Nsukka Formation in Ikpankwu Area. Petrol Coal 65(3)

Okeke KK (2017) Palynostratigraphy and granulometric assessment of Paleocene to early Miocene sediments. In: Awka–Onitsha Area, Niger Delta Basin, Southeastern Nigeria. MSc. thesis, University of Nigeria Nsukka, pp 1–240

Okeke KK (2023) Palynostratigraphy, palynofacies and paleoenvironment reconstruction of the Late Campanian to Danian strata of the Anambra Basin, Southeastern Nigeria [Ph.D Thesis]. University of Nigeria Nsukka, pp 1–344

Okeke KK (2024) Depositional facies and palynofacies provenance reconstruction of the Danian Nsukka Formation, Southeastern Nigeria. Arab J Geosci 17(5):157

Oloto IN (1990) Palynological assemblage from the Danian of South-West Nigeria. Acta Palaeobot 30(1)

Orife JM, Avbovbo AA (1981) Stratigraphic and unconformity traps in the Niger delta (abs). Am Assoc Pet Geol Bull 65:967

Ozkaya I (1988) A simple analysis of oil-induced fracturing in sedimentary rocks. Mar Pet Geol 5:293–297

Price LC (1989) Primary petroleum migration from shales with oxygen- rich organic matter. J Pet Geol 12:289–324

Rayment RA (1965) Aspects of the geology of Nigeria. Ibadan University Press, pp 2–115

Salami MB (1990) Palynomorph taxa from the "lower coal measures" deposits (? Campanian-Maastrichtian) of Anambra trough, Southwestern Nigeria. J Afr Earth Sci 11(1/2):135–150

Traverse A (1988) Plant evolution dances to a different beat: plant and animal evolutionary mechanisms compared. Hist Biol 1(4):277–301

Umeji OP (2005) Palynological study of the Okaba coal mine section in the Anambra Basin, Southern Nigeria. J Min Geol 41(2):194

Umeji OP (2006) Palynological evidence of the Turunian/Campanian boundary between the Abakaliki and the Anambra basins, as exposed at Leru along the Enugu-Port Harcourt expressway, southeastern Nigeria. J Min Geol 42(2):141–155

Umeji OP (2011) Palynofacies and palaeodepositional environment of Campano-Maastrichtian sediments exposed around Leru in Anambra Basin, southeastern Nigeria. J Min Geol 47(1):49–69

Umeji OP, Edet JJ (2008) Palynostratigraphy and paleoenvironments of the type area of Nsukka Formation of Anambra Basin, Southeastern Nigeria. Nig Assoc Petrol Explor Bull 20:72–89

Umeji OP, Nwajide CS (2007) Age control and designation of the standard stratotype of Nsukka Formation of Anambra Basin, southeastern Nigeria. J Min Geol 43(2):147–166

Umeji AC (2000) Evolution of the Abakaliki and the Anambra sedimentary basins, southeastern Nigeria. In: A report submitted to The Shell Petroleum Development Company Ltd, p 155

Umeji OP (2007) Late Albian to Campanian Palynostratigraphy of southeastern Nigerian Sedimentary Basins. Unpublished Ph.D Thesis, University of Nigeria Nsukka, Department of Geology 280 pp

Ungerer P (1990) State of the art of research in kinetic modelling of oil formation and expulsion. Org Geochem 16:1–26

Welte DH (1987) Migration of hydrocarbons. Facts and theory. In: Dolige B (ed) Migration of hydrocarbons in sedimentary basins. Editions Technip, Paris, pp 393–413

Zarboski PMP (1982) Campanian and Maastrichtian Sphenodiscid ammonite from Southeastern Nigeria. Bull Brit Mus Nat Hist 36:303–332

Zarboski PMP (1983) Campanian and Maastrichtian ammonite correlation and Paleogeography in Nigeria. J Afr Earth Sci 1:51–63

Chapter 2
Regional Geology and Basin Evolution

Abstract The petroliferous economic importance of the Anambra Basin highlights the significance of the basin forming tectonics and basin fills, substantiated by the eustatic sea level dimensions in sub Saharan African region depositional space that triggered several geologic formations and various economic resources inherent in the study. The unconformity system of the Anambra Basin reveals it as a distinct sedimentary basin overlying the Benue Trough as a separate lithostratigraphic domain of Late Campanian to Danian sedimentary units inherent in the Nkporo Group, Mamu, Ajali and Nsukka formations. The origin and tectonic events of the studied sedimentary basin were scientifically upheld as mantle plume model, plate tectonic model, the subduction model and pull apart origin tectonic evolution models by variable scholarly scientific debate to explain the evolution of the Anambra Basin and the older Benue Trough. In affirmation of geologic and scientific controversy of the Anambra Basin's origin, a tectonic evolution model was confirmed and the Anambra Basin was deemed as a separate basin resulting from the concomitant folding with the up-doming of the Abakaliki Basin (Benue Trough) and down-warping of the Anambra "platform' basin during the Santonian. The longitudinally faulted crust megatectonic structure, breakup of the West Gondwana supercontinent and the contemporaneous rifting of the Abakaliki domain and the Anambra Basin terrane sag that climaxed during the separation of south America from the African plate, illustrated as geologically accepted consensus and interpretation of the tectonic origin of the studied basin. The geologic debate and tectonic origin model quagmire of Anambra Basin highlights the principle of uniformitarianism as a geologic concept applied and updated with a series of research to buttress the scientific growth inherent in geology research.

Keywords Anambra Basin · Megatectonic structure · Gondwana supercontinent · Unconformity system · Stratigraphic facing

2.1 Geologic Framework

The geology of the Anambra Basin can be discussed under the origin and sedimentary basin fill. Many scholarly geoscientific interpretations and research on the origin of the Nigerian sedimentary basins led to the establishment of many hypothetical and scientific models to explain the evolution of the Anambra Basin (Fig. 2.1). However, with advancements in geologic scientific research and technology along with exposure to new geologic sections, it is now substantiated in this work and works of Umeji (2000), Nwajide (2013, 2022) and Whiteman (1982) that the Anambra Basin unconformably overlies the Benue Trough as a separate lithostratigraphic domain which consists sediments comprising the Nkporo Group, Mamu, Ajali and Nsukka formations. This substantiates the Anambra Basin as a distinct sedimentary basin previously documented by Umeji (2000), Nwajide (2013, 2022), Whiteman (1973, 1982), Murat (1972) and Ford (1981). The Anambra Basin consists of the Nkporo Group containing the Enugu Formation, Owelli Formation, Lafia Formation, Afikpo Formation, Otobi Formation as lateral equivalents; the Mamu Formation, Ajali Formation and Nsukka Formation.

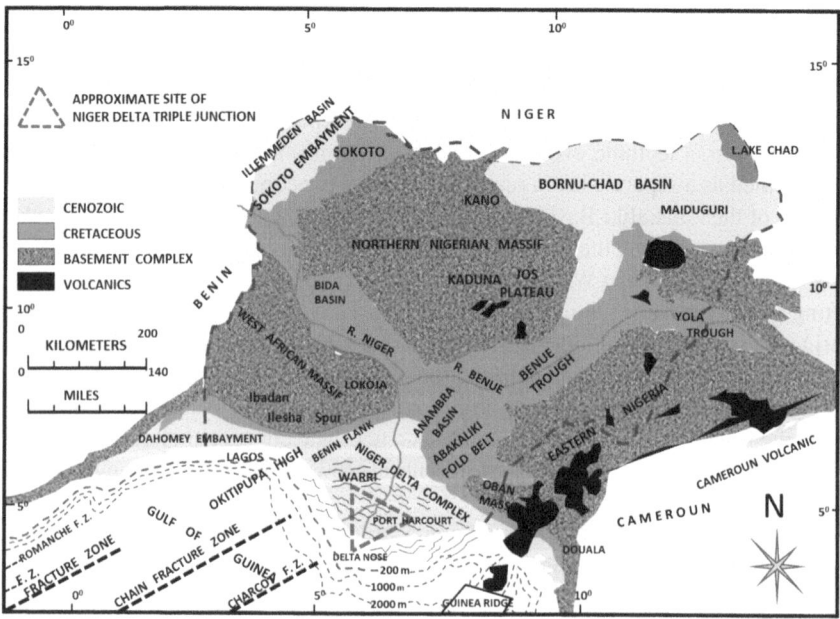

Fig. 2.1 Structural map of the Nigerian sedimentary basins showing the chain and charcot oceanic fracture zone (modified after Okeke 2023; Murat 1972)

2.2 Regional Tectonic Setting

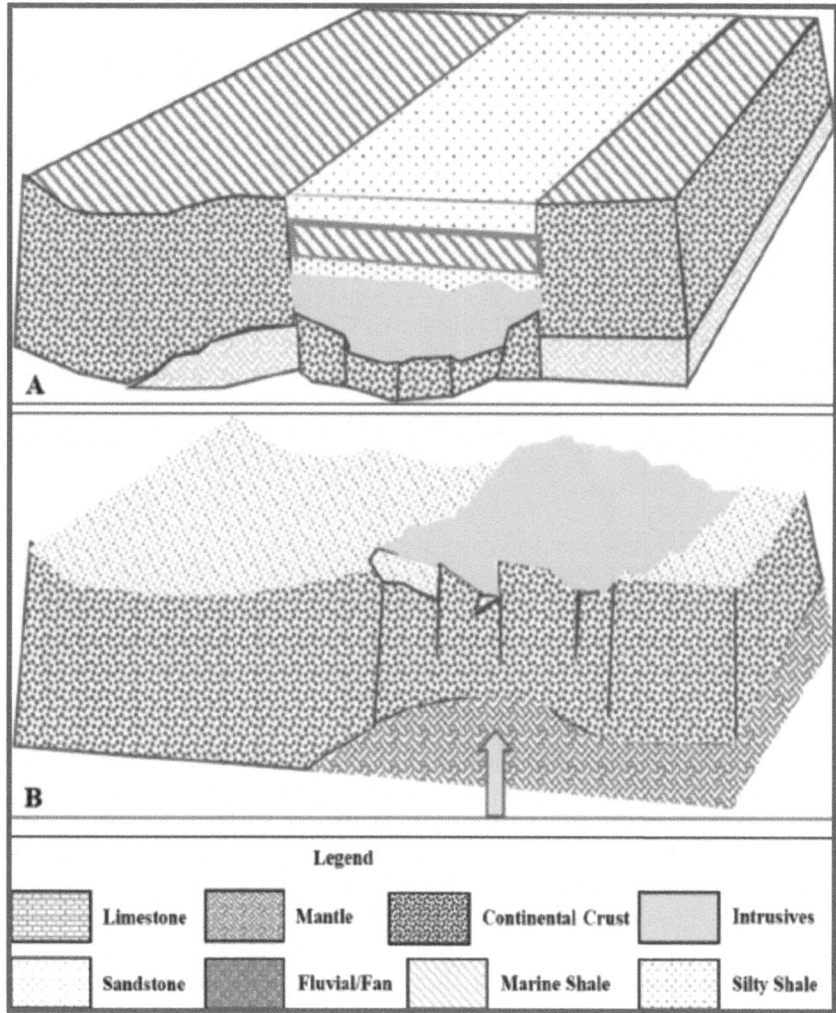

Fig. 2.2 a Filling of rift valley stage Early Cretaceous–Coniacian. **b** Post-rift intumescence stage uplift and folding showing the drowned Anambra Basin platform (Modified after Okeke 2023; Umeji 2000)

2.2 Regional Tectonic Setting

The Post Santonian origin of the Anambra Basin was variously discussed by different authors (Kogbe 1981; Benkhelil 1989, 1986; Nwajide 2013, 1990; Umeji 2000; King 1950, 1962; Whiteman 1982; Ojoh 1990) which led to many tectonic evolution models (cf Umeji 2000; Ojo 1990; Benkhelil 1989; Grant 1971; Burke et al. 1972; King, 1950, 1962; Wright 1968; Olade 1975) to explain the evolution of the Anambra

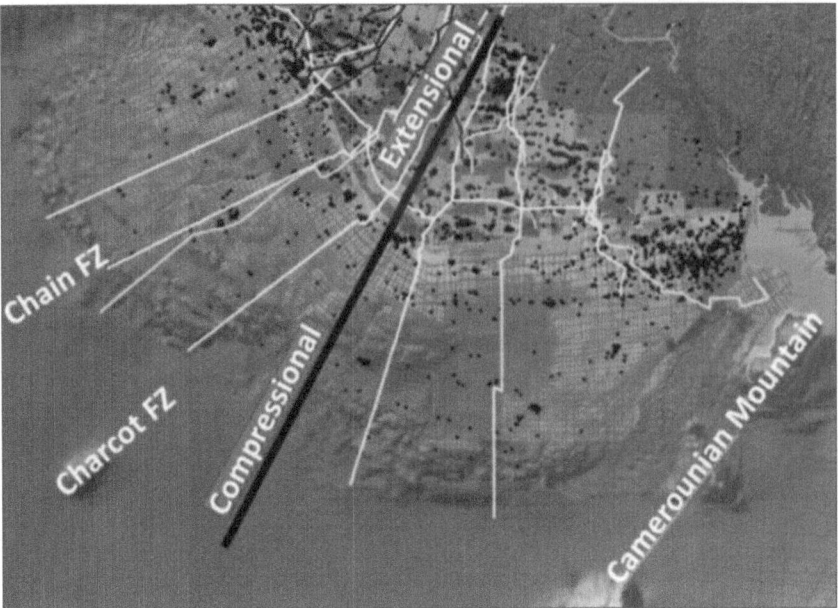

Fig. 2.3 Regional cross-section and Chain and Charcot lineaments of southeastern Nigeria (Nosike 2023)

Basin and the older Benue Trough (Fig. 2.2). The models are the mantle plume model (Olade 1975), plate tectonic model (Grant 1971; Burke et al. 1971, 1972), the subduction model (Nwachukwu 1972) and pull apart origin model of Benkheilil (1989). In some of these models the Anambra Basin have been erroneously regarded as an integral part of the Benue Trough and an alternative name for the Abakaliki Basin (i.e. Lower Benue Trough) e.g. Unomah and Ekweozor (1993), Kogbe (1981) and Nwajide (1990) while Nwajide (2013), Murat (1972), Umeji (2000), Whiteman (1982) and Ojoh (1990) deemed same as a separate basin resulting from the concomitant folding with the up-doming of the Abakaliki Basin (Benue Trough) and down-warping of the Anambra "platform' basin during the Santonian. According to these later proponent works (cf. Nwajide 2013; Murat 1972; Umeji 2000); the longitudinally faulted crust is the megatectonic structure that gave rise to the emergence of the Anambra Basin. The eastern half of the longitudinal fault subsided and became the southern Benue Trough (i.e. the Abakaliki sub-basin). However, Unrug (1996) argued that the breakup of the West Gondwana supercontinent during the Late Jurassic to Early Cretaceous times happened after the rifting of the Abakaliki (Benue Trough) Basin area and the saging of the Anambra Basin domain while Umeji (2000) opined that the rifting of the Abakaliki and the Anambra sag was contemporaneous to the separation of the south America from the African plate (Figs. 2.1 and 2.2). The contemporaneous rifting of the Abakaliki Basin and the Saging of the Anambra basin model of Nwajide (2013) confirmed that the boundary between the Anambra

2.2 Regional Tectonic Setting

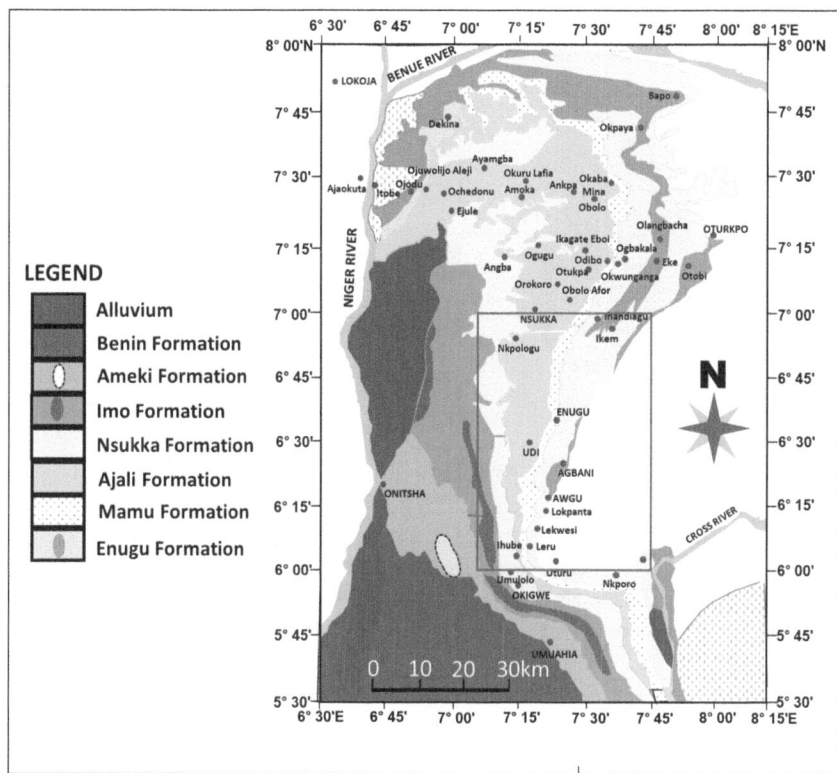

Fig. 2.4 Geological map of Nigeria showing Cretaceous successions of the Anambra Basin (modified after Umeji 2000; Okeke 2023)

and Abakaliki basins "was either drastically eroded or virtually removed during the 50 Ma, or nearest, that passed from the Early Cretaceous to the Campanian". This demarcates or separates the youngest formation of the Abakaliki Basin (Southern Benue Trough) the Awgu Formation from the Anambra Basin which has the Enugu Formation as the oldest stratigraphic unit/formation. The subsequent structural lows triggered sediment fills with alternating marine and continental successions, listric fault and hydrocarbon traps structural style (Nosike 2023). However, the seismic investigation of the Niger Delta Basin, Anambra Basin and Benue Trough, demonstrated a similar dynamic structural axis system, genetically related to the failed arm of the triple junction developed at the onset of the opening of the Atlantic Ocean (Fig. 2.2 and Table 2.1).

The tectonic events responsible for the evolution of the Anambra Basin and the stratigraphic framework of the Pre Santonian-Benue Trough were equally documented as the Pre Santonian-Albian, Santonian (Late Cretaceous) megatectonic framework and Post Santonian Campanian—Maastrichtian—Early Cenozoic megatectonic framework of Nwajide (2013), Umeji (2000) and Whiteman (1982). The

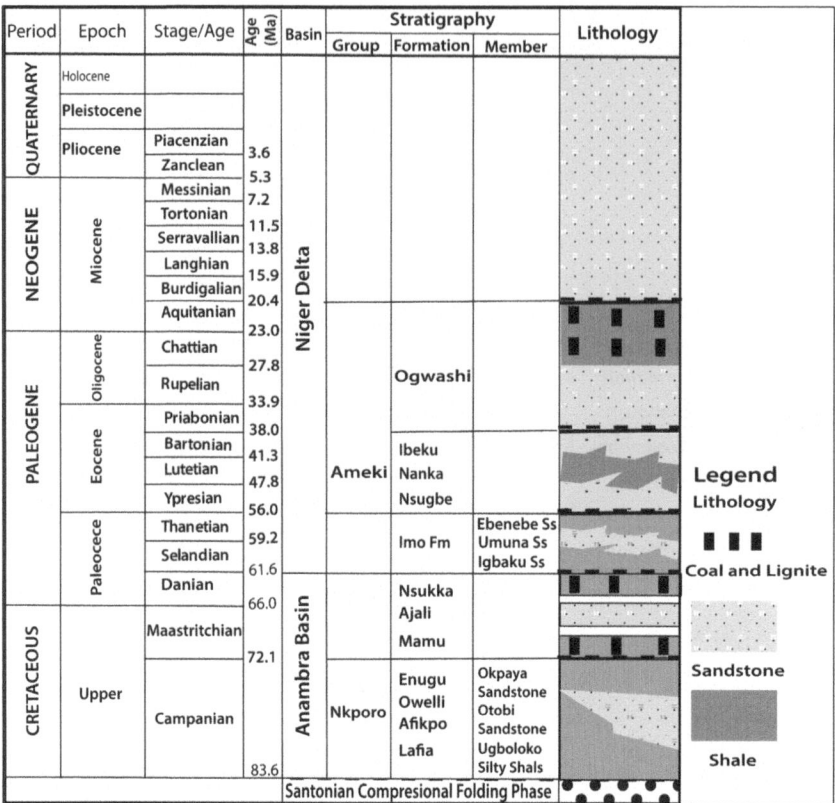

Fig. 2.5 Stratigraphic column of the Anambra and Niger Delta Basins, southeastern Nigerian (Okeke 2023)

tectonic folding of the Pre Santonian-Benue Trough sedimentary lithologies/strata and the unfolding and gentle tilting of the post Santonian upper Cretaceous strata were the integral demarcating geologic events of the Anambra Basin and the Benue Trough (Fig. 2.3). This indicates that the tectonic intrusion and folding event in the southeastern sedimentary archetypes of the Pre Santonian-Benue Trough, Post Santonian Anambra Basin domain and Nigeria, in general, was during the Santonian. This substantiates the Benue Trough and Anambra Basin folding and unfolding respectively. The Basement complex igneous and metamorphic rocks evolution and indices are of little importance in this sedimentary facies architecture and palynological interpretation of the successor basins. Although Basement complex igneous and metamorphic rocks are indicative of the robust, broad depositional architectures and mechanisms during the evolution of the Late Mesozoic and Cenozoic Post Santonian Anambra Basin inclined formations as documented by Nwajide (2013) and Umeji (2000). This confirms the dynamics of magmatic evolution and its role in the evolution of sedimentary basins. Generally, the Pre-Santonian tectonic and Post Santonian

2.2 Regional Tectonic Setting

Table 2.1 Regional geology, basin forming tectonics and basin fills of the Anambra basin terminology

Terminology	Definition
Stratigraphic facing	The direction to the youngest deposited sediment/bed of an outcropping section or the younging direction of the depositional top of beds.
Megatectonic structure	Megatectonics is a tectonics of very large structural geology features of the earth's crust within the continents and oceans consistent with the deformational structure of rocks in the earth's crust. It concerns megatectonic events of high magnitude faulting, folding and jointing.
Tectonics	Tectonics is a branch of geology concerned with structures of the crust that deals with all the deformational structures of rocks in the earth's crust. It pinpoints series of microtectonic structure domiciled in foliations, porphyroblasts, veins, and fringes.
Gondwana supercontinent	Gondwana was an ancient supercontinent that broke into Africa, South America, Australia, Antarctica, Zealandia, Arabia, and the Indian Subcontinent.
Chain and Charcot Fracture Zones	A fracture zone concerned with the separation of the Gondwana supercontinent.

framework of Whiteman (1982), Nwajide (2013), and Umeji (2000) report the unconformity or separating terms accurately which demarcates the two Mesozoic Basins of Nigerian southeastern domains based on the structural and lithological verities in the field (Fig. 2.3).

Notwithstanding the contemporaneous evolution of the two basins' architecture, the Anambra Basin sediments inflow was inactive while sediment accumulation was visible in the Abakaliki Basin/ Benue Trough from Albian to Turonian Times (Umeji 2007). After the tectonic up-doming of the Abakaliki Basin and saging of the Anambra Basin, the depressed Anambra Basin became the major depocenter for the deposition of marine and non-marine sediments. Analogies from other marginal basins such as the Dahomey Basin and those from other areas of the world including the continental shelves of West Africa and Brazil, indicated that the earliest sediment accumulation in Anambra Basin prior to the basin subsidence marine incursion was non-marine fluvial with lacustrine sandstones and shale strata (Umeji 2007). There are divergent opinions on the timing of these tectonic events and basin infill whereby Nwajide (2013) indicated Turonian or earlier age and Santonian to early Paleocene (Danian); while Umeji (2000) upheld an Early Cretaceous to Campanian along with Late Jurassic to Early Cretaceous age of Unrug (1996).

Although macropalontologic, palynological, foraminiferal micropaleontologic species occurrences indicated that the southeastern Nigerian sedimentary basins of lower Benue Trough and the Anambra Basin geologic sections depositional facies were of Aptian (Late Cretaceous) and -Danian (Early Paleocene) age. These facts substantiate that the tectonic uplift in the past in eastern Nigeria and adjacent Cameroun as well as the erosion of the Pre Santonian sedimentary rocks are important sources of clastic sediments of the Anambra Basin.

2.3 Regional Stratigraphy

The lithostratigraphic units of the Anambra Basin (Figs. 2.4 and 2.5) consists of the Nkporo Group comprising the Enugu Formation (Reyment 1965; Simpson 1954) with Owelli Sandstone (Reyment 1965) previously erroneously regarded as the Awgu sandstone (Simpson 1954), the Otobi Sandstone (Nwajide 2013; Umeji 2000; Whiteman 1982) along with the Okpaya siltstone and silty sandstone, Asata Formation (Agumanu 1993), along with the Ugboloko Silty shale sensu Umeji (2000). These lithostratigraphic formations of the Anambra Basin and those of the older Benue Trough units were documented by Ojo (1990) as the southeastern basin intermittent subsidence event packages which were higher in Pre Albian, low in the lower Cenomanian and extremely high during the Turonan. The intermittent subsidence timeframe was regarded as the important stage of platform subsidence during the evolution of the three southeastern Nigerian basins. Nwajide (2013) noted these timeframe episodic events as when Anambra Basin creation started in force in Coniacian stage and climaxed during the Santonian thermotectonic events. According to Ojoh (1990), the confined subsidence on the western extremities of the southern Benue Trough and continued sea level rise into Coniacian initiated the origin of the Anambra Basin. In line with this, weak zones and rifting phase structural attributes of the Cretaceous successions, the resulting structural lows triggered dynamic sedimentary rock fills with alternating marine and non-marine continental deposits, listric fault and array of hydrocarbon trapping style (Nosike 2023).

The Post Santonian upper Cretaceous triangular Anambra Basin were subdivided into zones A, B and C (Umeji 2000). Zone A encompasses the northern sector which comprises areas surrounding west of Oturkpo through Ojodu-Itobe axis on the Niger while Zone B stands for the Central Sector in the south of Nsukka-Ikem and Enugu areas. Zone C represents the Southernmost Sector which geographically extends from the South of the Nsukka-Ikem axis through Enugu vicinity and the regions south of Awgu to Leru and Okigwe, terminating in the vicinity of Uturu (Umeji 2000). These sedimentary piles and filling of the Anambra Basin with sedimentary rocks were done within 20 My time, from Santonian to Latest Maastrichtian. However, Nwajide (2013) indicated 60 My (i.e. Santonian to Early Paleocene) as the absolute time span of the depositional facies infill. The basin is bounded in the west by the Okitipupa ridge—a tectonically induced structure separating the Dahomey Basin and the southern Benue Trough. In contrast, around Auchi and Igara the basin overlies the basement northwesterly, and the sediment- basement contact is around Itobe and Anyigba (Nwajide 2013). The works of Nwajide (2013, 2022) noted that the boundary with the Bida Formation is an interfingering of Anambra Basin sediments and thus not well defined whereas the contact with the basement complex at Keffi and Wamba area is regarded as the Northern boundary.

In the southern domain, the onlap associated with the angular unconformity with the southern Benue Trough marks the southernmost boundary. The general thickness of the Anambra Basin is about 6 kilometres with the oldest formation overlying the folded Awgu Formation with an angular unconformity. Around the Lokpanta area the basin directly overlies the Ezeaku Formation (Umeji 2000), while at the Uturu section, it overlies the Asu River lithostratigraphic units. The basin was also regarded as the thickest marginal to the fold belts, with an estimated 30000 ft (Whiteman 1982). Generally, in the Calabar and the Benin Flanks, sedimentologic, biostratigraphy, micropaleontology and sequence stratigraphic evidence have shown that the Enugu Formation of the Nkporo Group is the only deposited lithostratigraphic unit of the Anambra Basin. In contrast, the conspicuous younger Mamu, Ajali and Nsukka outcropping basin units at the southeastern and northern section were not deposited.

References

Agumanu AE (1993) Sedimentology of Owelli sandstone (Campano-Maastrichtian) Southern Benue Trough Nigeria. J Mining Geol 29(2):21–35
Benkhelil J (1989) The origin and evolution of the Cretaceous Benue Trough (Nigeria). J Afr Earth Sc 8(251–282):320
Benkhelil J (1986) Caracteristiques structurales et evolution geodynamique du basin intracontinentale de la Benoue (Nigeria). Thesis d'etat, Nice, 275 pp
Burke KC, Dessauvagie TFJ, Whiteman AJ (1971) The opening of the Gulf of Guinea and the geological history of the Benue depression and Niger delta. Nat Phys Sci 233:51–55
Burke KC, Dessauvagie TFJ, Whiteman AJ (1972) Geological history of the Benue valley and adjacent areas. In: Dessauvagie TFJ, Whiteman AJ (eds) African geology. University of Ibadan Press, Nigeria, pp 187–218
Ford SO (1981) The economic mineral resources of The Benue Trough. Earth Evol Sci V1(2):154–163
Grant NK (1971) South Atlantic, Benue Trough, and Gulf of Guinea Cretaceous triple junction. Geol Soc Am Bull 82(8):2295–2298
King LC (1950) Outline and disruption of Gondwanaland. Geol Mag 87:353–359
King LC (1962) Morphology of the earth; a study and synthesis of world scenery. 69–70. Oliver & Boyd
Kogbe CA (1981) Cretaceous and tertiary of the Iullemmeden basin in Nigeria (West Africa). Cretac Res 2(2):129–186
Murat RC (1972) Stratigraphy and paleogeography of the Cretaceous and lower tertiary in southern Nigeria. In: Dessauvagie TFJ, Whiteman AJ (eds) African geology. University of Ibadan Press, Ibadan, pp 251–266
Nosike L (2023) Impact of structural dynamics on hydrocarbon of the deposits in the Enugu axis of the Cretaceous Anambra Basin. PTDJ, July vol 13, no 2
Nwachukwu SO (1972) The tectonic evolution of the southern portion of the Benue Trough, Nigeria. Geol Mag 109:411–419
Nwajide CS (1990) Cretaceous sedimentation and palaeogeography of the Central Benue Trough. In: Ofoegbu CO (ed) The Benue Trough structure and evolution. Friedr. Vieweg and Sohn, Braunchweig/Wiesbaden, pp 19–38
Nwajide CS (2013) Geology of Nigeria's sedimentary basins. CSS Bookshops Ltd., pp 347–411
Nwajide CS (2022) Geology of Nigeria's sedimentary basins. 2nd ed. Albishara Educational Publications, pp 394–443
Ojoh K (1990) Cretaceous geodynamic evolution of the southern part of the Benue Trough (Nigeria) in the equatorial domain of the south Atlantic: stratigraphy, basin analysis and paleogeography. Bull Centres Rech Explor Prod Elf Aquitaine 14:419–442

Okeke KK (2023) Palynostratigraphy, palynofacies and paleoenvironment reconstruction of the Late Campanian to Danian strata of the Anambra Basin, Southeastern Nigeria [Ph.D Thesis]. University of Nigeria Nsukka, pp 1–344

Olade MA (1975) Evolution of Nigeria's Benue trough (aulacogen): a tectonic model. Geol Mag 112(6):575–583

Rayment RA (1965) Aspects of the geology of Nigeria. Ibadan University Press, pp 2–115

Simpson A (1954) The Nigerian coal field: the geology of parts of Owerri and Benue Provinces. Geol Surv Niger Bull 24:1–85

Umeji AC (2000) Evolution of the Abakaliki and the Anambra sedimentary basins, southeastern Nigeria. In: A report Submitted to The Shell Petroleum Development Company Ltd, p 155

Umeji OP (2007) Late Albian to Campanian Palynostratigraphy of southeastern Nigerian Sedimentary Basins. Unpublished Ph.D Thesis, University of Nigeria Nsukka, Department of Geology, 280 pp

Unomah GI, Ekweozor CM (1993) Petroleum source-rock assessment of the Campania Nkporo Shale, Lower Benue Trough, Nigeria. NAPE Bull 8(2):172–186

Unrug R (1996) The assembly of Gondwanaland. Episodes 19:11–20

Whiteman A (1982) Nigeria: its petroleum geology, resources and potential, vol 2. Graham and Trotman, London, p 394

Whiteman A (1973) Geology and hydrocarbon prospects of Nigeria, vols 1 and 2, pp 1–257 (with 76 Figs and 38 Tables). Exploration Consultant Limited Marlowe, England (Compiled 1971 and 1972)

Wright JB (1968) South Atlantic continental drift and the Benue Trough. Tectonophysics 6(4):301–310

Chapter 3
Outcrop-Based Studies and Lithologic Description of Outcropping Stratigraphic Successions

Abstract Six major sedimentary rock outcropping lithofacies units allowed a detailed comprehension of the hydrodynamics and sedimentation of the studied sedimentary rock units of the Anambra Basin. They comprise limestone, siltstone, mudstone, shale, and variable sandstone grain size parameters. The rocks were studied and logged for a holistic appraisal and comprehension of the tidal equilibrium and dynamic tidal theory (model). This demonstrates effects of current, waves and quite water mechanics relationship and other hydrodynamic intricacies inherent in the Anambra Basin during sediment deposition. Several outcrops at spring watersheds, rivers, road cuts, gullies and quarry sites were exposed. They present a relative NNE-SSW strike line and a dipping inclination of NNW direction with an average dip amount of 6°. The studied shale lithologies are black to grey colored, carbonaceous, faulted, and interlaminated with open and closed fractures. The facies are intermittently overlain and underlain by bioturbated, plane planar laminated and planar crossbeded sandstone. It displays a systematic vertical and lateral facies change of the sandstone layers strengthened in the stacked heterolithic and parallel laminated siltstone and shale with an erosional base significant of rip-up clasts. The joint orientation trends NW-SE and NE-SW with paleocurrent measurements highlights of the diverse sedimentary structural trap dynamics of the footwall and hanging wall of a normal fault along with a series of open and closed joints structural elements of the studied outcrops. The research in this chapter upholds the dynamics of the geologic concept of 'time and space' in the deposition and structural reorientation of deposited sediments through tectonics build-up activities on the earth's crust.

Keywords Depositional contacts · Outcrops · Compacted beds · Footwall · Hanging wall · Massive beds

3.1 Field Description

Sedimentary rocks outcrop studies show that several sandstone ridges and valleys characterize the studied section of the Anambra Basin typified by the Enugu Formation of the Nkporo Group and the Danian units of the Nsukka Formation of this study (Fig. 1.1 and 3.1). Several outcrop sections were studied in this cause of the field investigation but five (5) major outcrops were described in this essay due to the structural dimensions of the outcrops (Table 3.1). These structural styles of the formations are inherent in sedimentary rock key structures footwall, hanging-wall and anticlinal structures with series of closed and open joint structures.

The prominent sandstone outcrop ridges include those of Aguabo pedestrian bridge, Neke Isi-Uzo road cut ridge, Ozalla ridge and Ikpankwu quarry ridges (Fig. 3.1). Sand bodies studied in the area are discontinuous due to natural and human activities of erosion and urbanization (less than 1 km to few kilometers in length) and trends in E-W and NNE-SSW directions. Quarrying activities and road construction on these sandstone ridges provided well exposed sections for outcrop studies.

Fig. 3.1 Geomorphological view of series of ridges and scarp face of the Enugu cuesta in the vicinity of Enugu, Enugu Formation

3.1 Field Description

Table 3.1 Synoptic description of some key sedimentary rock stratification terminology of the study

Terminology	Definition
Depositional contacts	A sediment layer/bed deposited over preexisting rock sedimentary rock type with a definite bedding plane
Bed	A deposited sedimentary rock primary layer, characterized by changes in degree of composition, texture, colour and sedimentary structures defined by bedding plane or sedimentary rock parting in outcropping sediments
Compacted beds	Sedimentary rock layers lithified by pressure on unlithified strata by the weight of the overlying younger sediments
Massive beds	A sedimentary rock deposited beds without any physical internal layering and sedimentary structures

The field mapping and laboratory empirical analysis indicated that understanding the sediment transportation processes and deposition is crucial in the deposited sediments' hydrodynamics and sedimentary structural interpretation. The described outcrop sections logged across the vicinity are summarized in Tables discussed below.

3.1.1 Aguabo Pedestrian Bridge, Ridge Section (Outcrop 1)

This section is exposed on the Aguabo pedestrian bridge road cut opposite Gulf estate Enugu through road construction activities along the Enugu-Onitsha express (Figs. 3.2 and 3.3). The Aguabo pedestrian bridge road cut section comprises stacked sediment thickness of about 34 m. Lithologic units of the study are made up of predominantly grey to dark carbonaceous shales and sandstones extending from the Aguabo pedestrian bridge to Iyi Obune and Mmiri Ugwu Onyeama confluence river channel with a pronounced disconformity depositional contacts model. The lithofacies units of the Aguabo pedestrian bridge section consists of siltstones, mudrock, dark-grey carbonaceous shales and series of consolidated sandstone beds with pronounced lateral facies changes because of the growth fault structural style in the study areas (Fig. 3.2).

The pronounced black carbonaceous faulted shale outcrop was interbedded by a 5 m heterolithic bed and interlaminated spasmodically by siltstone bands of 20–30 cm thick (Fig. 3.3). The stacked shale strata grade into heteroliths of interbedded fine-grained sandstones and siltstone. The prominent features observed in this section are the syndepositional listric normal fault and growth fault system along with a series of closed and open joint structures (Fig. 3.2).

Fig. 3.2 The Enugu Formation outcrop section at Aguabo pedestrian bridge road cut along Enugu-Onitsha road, Enugu. **b** Carbonaceous shale of the Enugu Formation exhibiting the infilled fractures. **c** Heteroliths of carbonaceous gray shale and siltstone with infilled fractures of the Enugu Formation at Enugu

3.1.2 Neke Isi-Uzo Road Cut Ridge (Outcrop 2)

This section is exposed on the Neke Isi-Uzo ridge in the Ugwuogo vicinity off Opi-Abakpa-Nike bypass through road construction activities along Ugwuogo Neke-Ikem Road. The Neke Isi-Uzo ridge road cut section comprises a stacked sediment thickness of about 15.5 m. Lithologic attributes of the sedimentary outcrop consists of well-preserved grey siltstones, shales and sandstone facies. The logged section

3.1 Field Description

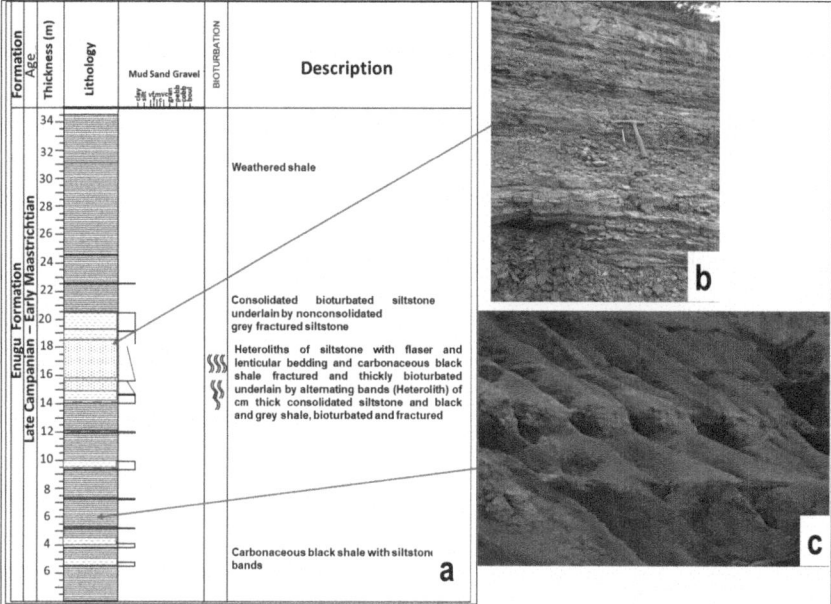

Fig. 3.3 **a** Litholog of the Aguabo section of the Enugu Formation. **b** Heteroliths of carbonaceous gray shale and siltstone with infilled fractures of the Enugu Formation at Enugu area. **c** The Enugu Formation outcrop section at Aguabo pedestrian bridge, Enugu

revealed that outcrop sediments in the study are dominated by well-preserved dark carbonaceous shales with siltstone laminations that gradually thickens from 4 to 10 cm thick siltstones and kaolinitic shale bands. The lower section of the outcrop is made up of dark grey shale units with some relatively thin, well-stratified sandstone and siltstone of varying thicknesses. The shale facies are spasmodically overlain and underlain by stacked consolidated planar laminated sandstone, siltstone with sequence of closed and open joints in some places. Lateral facies change outline of the stacked strata is depicted as heterolithic, parallel laminated silts and shales. The obvious erosional bases of the sequences have diagnostic rip up clasts and closed and open joints. The noticeable sedimentary structural attributes of the outcropping Enugu Formation sediments observed in this section are a series of conspicuous syndepositional listric normal faults, conjugate arrays, synthetic and antithetic faulted bed rock outcropping sequences (Table 3.2). Several sharp and gradational depositional contacts with a relative NNE-SSW strike line and dipping inclination of NNW direction with an average dip amount of 6° (Fig. 1.1) were generated in the field exploration cause.

Table 3.2 Summary of outcrop locations in the Aguayo pedestrian bridge ridge section Enugu, Neke Isi-Uzo road cut ridge at Neke-Isiuzo Ugwuogo vicinity and Amagu section by Enugu-Port Harcourt roadcut, Enugu

Aguayo pedestrian bridge, ridge section (outcrop)	
Outcrop specifics	Latitude N 06° 28' 09" and Longitude E 007° 28' 17.3"
Region/Province	Enugu, southeastern Nigeria
Formation	Enugu
Chronostratigraphy/age	Late Campanian–Early Maastrichtian
Thickness	34.5m (113.189 Ft)
Lithology	Carbonaceous shale, siltstone, kaolinitic shale bands, fine-grained sandstone, consolidated sandstone and heteroliths
Structural imprints	Syndepositional listric normal fault and growth fault, open and closed joint structures, parallel lamination, intra-formational clast
Neke Isi-Uzo road cut ridge (Outcrop 2)	
Outcrop specifics	Latitude N 06° 43' 32" and Longitude E 007° 38' 42"
Region/Province	Neke Isiuzo, southeastern Nigeria
Formation	Enugu
Chronostratigraphy/age	Late Campanian–Early Maastrichtian
Thickness	15.5 m (50.85302 Ft)
Lithology	Carbonaceous shale, siltstone, fine-grained sandstone, consolidated sandstone; kaolinitic shale bands and heteroliths
Structural imprints	Open and closed joint structures, siltstone lamination, intra-formational clast; well-stratified sandstone and siltstone, planar laminated sandstone, parallel laminated silts and shales, syndepositional listric normal faults, conjugate arrays, synthetic and antithetic faulted bedrock
Amagu section by Enugu-Port Harcourt roadcut (outcrop 3)	
Outcrop specifics	Latitude N 06° 20' 57.7" and Longitude E 007° 29' 23.6"
Region/Province	Enugu, southeastern Nigeria
Formation	Enugu
Chronostratigraphy/age	Late Campanian–Early Maastrichtian
Thickness	15.3 m (50.19685 Ft)
Lithology	Carbonaceous shale, siltstone, medium–fine-grained sandstone, consolidated sandstone, kaolinitic shale bands and heteroliths
Structural imprints	Open and closed joint structures, siltstone lamination, intra-formational clast, well-stratified sandstone and siltstone, planar laminated sandstone, parallel laminated silts and shales; normal faults

3.1 Field Description

Table 3.3 Summary of outcrop locations in the Ozalla/Four-Corner road cut section and the Ikpankwu ridge quarry section

Ozalla/four-corner road cut (outcrop 4)	
Outcrop specifics	Latitude N 06° 19' 07.5" and Longitude E 007° 28' 77.6"
Region/Province	Enugu, southeastern Nigeria
Formation	Enugu
Chronostratigraphy/age	Late Campanian– Early Maastrichtian
Thickness	7.5 m (24.6063 Ft)
Lithology	Carbonaceous shale, siltstone, consolidated sandstone, mottled clay and kaolinitic shale bands
Structural imprints	Open and closed joint structures, siltstone lamination, stratified sandstone and siltstone, parallel laminated silts and shales; normal faults
Ikpankwu ridge section (outcrop 5)	
Outcrop specifics	Latitude N 05° 51' 17.8" and Longitude E 007° 21' 21.9"
Region/Province	Ikpankwu area, Okigwe southeastern Nigeria
Chronostratigraphy/age	Danian
Thickness	27 m (88.5827 Ft)
Lithology	Carbonaceous shale, siltstone, consolidated sandstone, mottled clay and kaolinitic shale bands, heteroliths, mud rocks, medium—fine grained consolidated sandstone, bioturbated
Structural imprints	Open and closed joint structures, siltstone lamination and bands, stratified sandstone and siltstone, parallel laminated silts and shales; normal faults, erosional base, rip up clasts

3.1.3 Amagu Section by Enugu-Port Harcourt Roadcut (Outcrop 3)

This sub ridge section is exposed as a road cut along Enugu-Port Harcourt road, Enugu with a lateral 5 km that extends from the Amagu section to Ozara/Four-Corner junction section and Ihe Agbaogugu junction before intermixing with the older Benue Trough sediments and imprints of the Anambra Basin strata. Outcrops 3 is exposed within a sub-ridge. Within this sub-ridge exists a famous fault plane designated as the Amagu fault. Logged outcropped sections revealed that the basal heterolithic facies and carbonaceous mudrock facies dominate the lithologic units of the outcrop made up of grey to dark muds and shale units with intermittent relatively thick, well-stratified siltstone and medium- to very fine-grained sandstone beds of varying thicknesses (Fig. 3.4d and Table 3.2).

The stacked shale strata are regularly interbedded with relatively ferruginized siltstones and ironstone units that grade into heterolithic strata of about 1.5 m thick at the Amagu section (Fig. 3.4 d). The depositional contact of the stratified stacked strata contacts are sharp and strikes NNE-SSW, along with a measured NNW dip

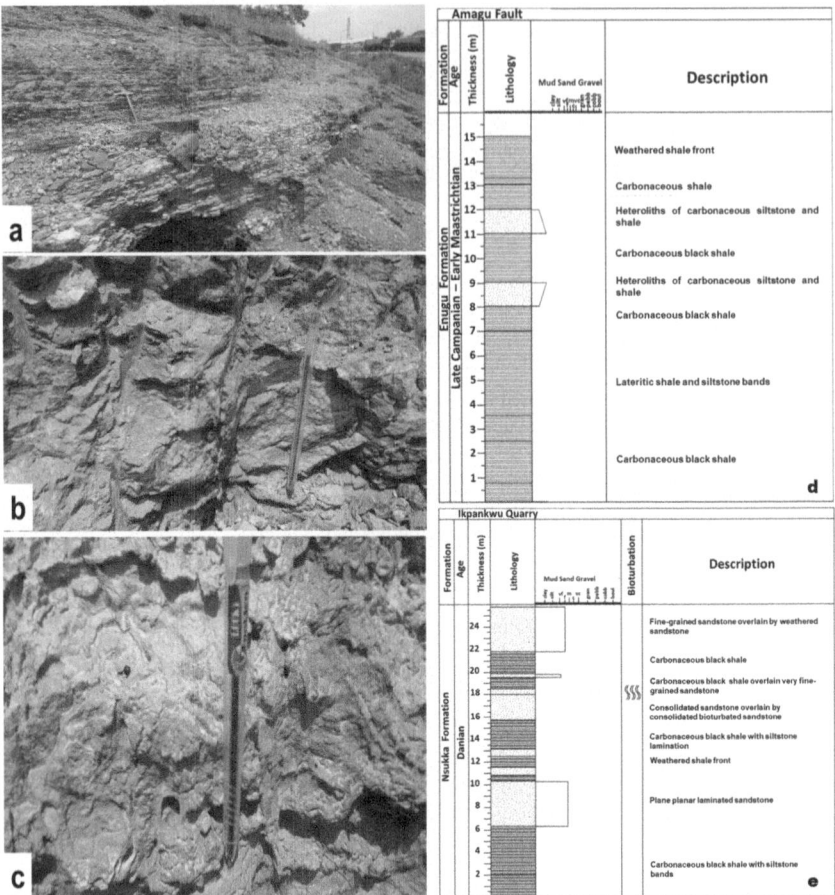

Fig. 3.4 a Heteroliths of carbonaceous gray shale and siltstone underlain by gray carbonaceous shale of the Enugu Formation at Umeonu flyover Enugu. b Infilled fractured siltstone of the Enugu Formation. c Close up on the infilled fractured siltstone of the Enugu. d Measured sedimentologic log of the Amagu Fault section of the Enugu Formation in the Enugu vicinity. e Measured sedimentologic log of the Ikpankwu quarry section of the Danian outcrop in the vicinity of Okigwe

direction with an average inclined dip amount of 6° (Fig. 1.1). Well-developed joint structures with infilling of brown colored sediments, systematically juxtaposed siltstone/sandstone block on shale block, hanging-wall and footwall of normal fault system characterized the exposed rock units of the sections in this ridge.

Fig. 3.5 a Carbonaceous black shale overlain by fairly consolidated planar cross bedded sandstone with contact between the shale and sandstone of the Danian outcrop section at Ihube. **b** Normal contact between carbonaceous black shale and the sandstone unit of the Danian section of the study at Ikpankwu quarry

3.1.4 Ozalla/Four-Corner Road Cut (Outcrop 4)

This outcrop section is located in the vicinity of Ozalla's four corner junction along Enugu—Port Harcourt Road. This key outcrop is exposed within a sub-ridge. Within this sub-ridge exists a key fault plane prototype designated as the Ozalla/Four-Corner fault. This section belongs to the Enugu Formation. The studied outcropped section shows that the lithologic units of the outcrop are dominated by basal carbonaceous shale facies interbedded by ferruginised sandstone which is similar to ironstone. The section is generally about 7.5 m thick dark carbonaceous shale inter-bedded with siltstone and mottled clay. This section is faulted and jointed (Table 3.3). The strata strike NNE-SSW, along with a measured NNW dip direction with an average inclined dip amount of $6°$ (Fig. 1.1). The fault structural trap here is a strati-structural or combination structural trap style pinpointing normal fault with a throw of 30 cm and heave 25 cm, along with series of open and closed joints structural elements of the outcrop. The joint orientation here trends NW-SE and NE-SW. Wedge-like and pinch-out structures were also encountered at the basal unit the studied Enugu Formation outcrop.

3.1.5 Ikpankwu Ridge Section (Outcrop 5)

This section is exposed at the Ikpankwu quarry with a stacked sediment thickness of 27 m and a thick cultivatable overburden situated at a relatively 2 km off the Enugu—Port Harcourt road. The Ikpankwu ridge section is a sedimentary rock deposit of the Danian unit of the Nsukka Formation, the youngest outcropping unit of the Anambra Basin. The section comprises thick sandstone strata consisting of siltstone, mud rocks, dark grey shale and sandstone lithologic units (Fig. 3.5 and Table 3.3). The basal bed is made up of carbonaceous shale with siltstone bands of 5–10 cm and varying thickness of laminations (Fig. 3.4e). This shale unit of the formation is spasmodically overlain by plane planar laminated and planar crossbeded sandstone along with a series of consolidated sandstone that is bioturbated in some units of the outcrop (Fig. 3.5). The lateral facies change of the sandstone layers in the formation is strengthened in the stacked heterolithic, parallel laminated siltstone and shale with an erosional base that is significant of rip up clasts. The evident sedimentary structural trap dynamics of the outcrop ridge section includes the footwall and hanging wall of a normal fault along with a series of open and closed joints structural elements of the outcrop.

3.2 Synopsis of the Lithologic Units Field Physiognomy and Facies Change Structural Trap Paradigm

The effect of structural deformation is noticeable in the structural trapping system, and lateral facies changes are evident in the stratigraphic succession of variable juxtaposed siltstone and sandstone blocks on the shale block. The distribution of bedding plane and sedimentary rock succession in many outcrop locations substantiated that the measured and described stratigraphic succession juxtaposed each other along the stacked fault plane. In contrast, the unfaulted layers overlap each other conformably when stacked together (Fig. 3.2).

The key structural trap style studied in the Enugu Formation are the Amagu fault Enugu along Enugu-Port Harcourt express road, Neke Isi-Uzo road cut ridge along Ugwuogo Neke-Ikem road, Aguabo fault within Aguabo pedestrian bridge Enugu, along Enugu-Onitsha Road and the Ozalla fault within Four-Corner junction along Enugu-Port Harcourt road. This structural trap style is marked with fault strikes $51°$ azimuth and a relevant inclined dip $53°$ with dip direction N321°W as obtained from the symbolic Aguabo Fault zone common to various fault zones of the studied formation. The Ikpankwu fault is an illustrative structural trap style system of the Nsukka Formation Danian domain. The conspicuous syndepositional listric normal faults, conjugate arrays, synthetic and antithetic faulted bed rock outcropping sequences, and open and closed joints are the structural trap style architecture of the formations of the Anambra Basin. The measured attitude of observed beds indicates that stratigraphic packages trend northeast to southwest (NNE-SSW) and dip in a NNW direction with a varying gauged range of dip amounts of $5°$-$6°$. The sedimentary rock profile of shale, siltstone, and sandstone are the characteristic outcropping lithologic rock type predominant in the study area with a measured variable thickness of as low as 0.2 m to as high as 6 m within weathered and unweathered sections similar to the lithofacies units of the Nsukka Formation under study (Figs. 3.2, 3.3, 3.4 and 3.5).

Primary sedimentary structures mechanics of parallel lamination and tabular cross-beds and secondary sedimentary structures of faults and joints dynamics encountered on the sandstone units are instrumental in the apprehending depositional orientation mechanics of the lithologic units (Figs. 3.2, 3.3, 3.4 and 3.5). The key structural distinctive parts of rocks that are evident in the studied vicinity are the conjugate arrays, synthetic and antithetic faulted rocks, and syndepositional listric normal faults encountered in the studied Enugu and Nsukka formations section at Neke Isi-Uzo, Enugu through Ikpankwu vicinity sections. These joints and fault structures are products of tectonic events typical of post-depositional events synonymous with secondary events. The synthesized geologic map graphics, depict the various outcrop locations and the associated equivalent geology formation displayed as outcrop 1–9, and others shown in Fig. 1.1 for the Enugu Formation. In contrast, outcrop 8-e stands for the Nsukka Formation while all were exposed within the Anambra Basin domain.

Chapter 4
Palynofacies Depositional Environment, and Source Rock Potential of the Enugu Formation and the Danian Outcropping Lithostratigraphic Units

Abstract The synthesized palynofacies origin and depositional environment of particulate organic matter demonstrate the geologic concept of 'space' in the origin and deposition of sedimentary rocks and igneous pedogenesis. The quality and quantity of kerogen particles are products of depositional environment theorized and practicalized in the concept of 'space' in defining the origin (evolution) of geologic processes. The space concept upholds the abundance and hydrodynamics quality of the visual palynofacies elements substantiated by the abundance of relatively well-preserved yellow and dark brown amorphous organic matter, marine taxa, opaque particles, with few dark brown structured phytoclasts and terrestrial palynomorphs. The palynofacies components illustrates the dynamic paleoenvironment of transitional marine settings of the inner neritic zone with intermittent outer neritic influences consistent with estuarine, nearshore and freshwater environments based on the normal varied quality and quantity of palynofacies constituents and profusion of land-derived microflora over marine dinoflagellate cysts for the Enugu Formation. The depositional array of the Danian lithostratigraphic unit sediments indicates shallow marine settings with deep marine influenced macro environments, embedded as products of lower and upper deltaic plains with similar variable palynofacies standard inherent in the Enugu Formation. Palynofacies elements suggest a thermally mature Type II–III kerogen indicative of oil and gas-prone hydrocarbons for the formations. The kerogen types are conceptualized in the oil seepages and natural gas flare within the Anambra Basin.

Keywords Palynofacies · Phytoclast · Kerogen Type · Palynomorph · Depositional environment

4.1 Origin and Palynofacies Constituents Types of the Enugu Formation and the Danian Lithostratigraphic Units

Kerogen is an organic element that is insoluble in organic solvents, water and other oxidizing acids whereas the natural fraction of the organic material soluble in organic solvents is regarded as bitumen which stands for oil in a solid state. Kerogen (Combaz 1964), palynofacies (Tyson 1993, 1995), and palynomaceral (Whitaker 1982, 1984) were independently proposed for definition and classification of organic matter. The various definitions and classification schemes result from different palynological techniques and specific scope of each palynofacies examination. In this work, the scheme of Tyson (1993, 1995) and particulate organic matter definition uplift which defined palynofacies in a practical sense as the quantitative and qualitative palynological study of the total particulate organic matter present in a sedimentary rock, was used. The four main groups were identified by Combaz (1964), Tyson (1993, 1995), Ibrahim et al. (1997), Okeke and Umeji (2018a, b), Okeke et al. (2023a, b) Okeke (2023) and Okeke (2017, 2023) and this work (Fig. 4.1) which include:

Phytoclasts: All structured and unstructured dispersed clay to fine-grained sandstone size particles.

Palynomorphs: All acid resistant organic-walled microscopic organisms.

Amorphous Organic Matter (AOM): All terrestrial and marine fluffy to dark brown dense structureless dispersed plant-derived kerogen particles.

Opaque Particles: Structured and unstructured brownish-black to black carbonized plant-derived kerogen particles.

These main group classification schemes and descriptions of all palynofacies components bring to light the hydrodynamic perspective, taphonomy, quantity and quality of the amorphous organic matter, opaque particles, structured and unstructured phytoclast, and resin. Other kerogen plant remains particles of angiospermous and gymnospermous vascular and non-vascular plant, palynomorphs of marine and non-marine origin and other marine particulate organic matter that remain substantiated in this research. These particulate organic matter groups' size ranges and origins are concurrently analogues with sedimentary grain size dynamics and sedimentary structures of shale, mudstone and siltstone facies. These palynofacies main groups were characterized with respect to the palynofacies prototype of Combaz (1964), Tyson (1993, 1995), Ibrahim et al. (1997), Okeke and Umeji (2018a, b), Okeke et al. (2023a, b) and Okeke (2017, 2023) exhibiting the same evolutionary, hydrodynamic, environmental and taphonomy relationship as follows (Table 4.1).

4.1 Origin and Palynofacies Constituents Types of the Enugu Formation …

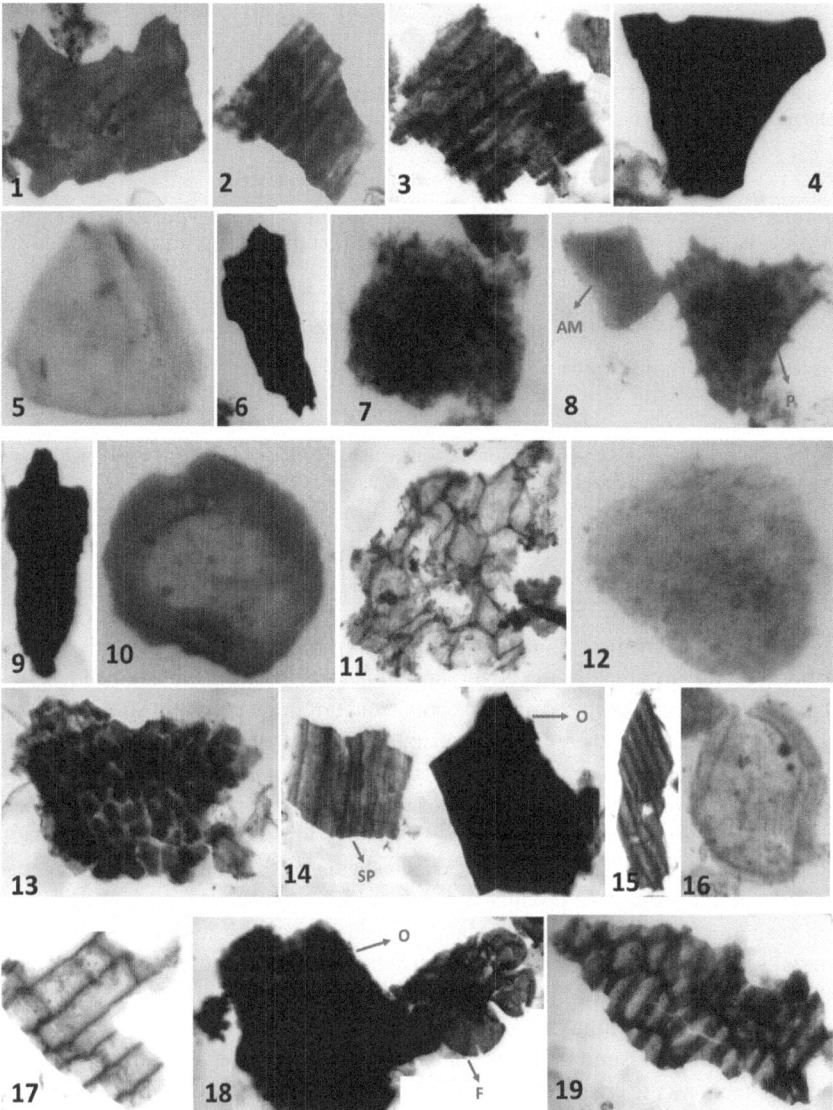

Fig. 4.1 Photomicrographs of particulate organic matter elements in the Enugu Formation; *O* opaque debris, *SP* structured phytoclast, *C* cuticle, *P* pollen, *AM* amorphous organic mass, *F* Foraminifera test lining. 1. Resin 2. Well preserved structured phytoclasts 3. Well preserved structured phytoclasts 4. Equant opaque debris 5. *Proteacidites sigalii* Boltenhagen 1978 6. Lath shaped opaque particle 7. Amorphous organic matter 8. Amorphous organic matter and *Echitriporites trianguliformis* Form B Lawal and Maullade 1986 9. Lath shaped opaque particle 10. *Cingulatisporites ornatus* Jan du Chêne et al. 1978 11. Partly degraded Cuticle 12. Amorphous organic matter 13. Degraded black brown phytoclast 14. Particulate organic matter components 15. Well-structured phytoclasts 16. Dinoflagellate cyst 17. Cuticle 18. Particulate organic matter components and 19. Cuticle

Table 4.1 The meaning of some outlined key palynofacies terminology

Terminology	Definition
Amorphous Organic Matter	Amorphous organic matter is a product of bacterial alteration of plants predominant in coastal and brackish settings. It occur as fluffy, yellow-amber and brown structureless dense substances
Resin	Resin particles are yellowish to amber coloured stem tissues of plants
Unstructured Phytoclasts	A degraded, irregularly shaped structureless to partly structured phytoclasts with brown to dark brown colour range without any observable cell structure
Cuticle Phytoclast	Cuticles are flat, platy, relatively thin phytoclast with well defined outline of epidermal cells along with distinct upper and lower epidermis
Structured (wood) Phytoclasts	Wood remains of plants of terrestrial origin
Opaque Debris	Opaque particles are products of oxidation of translucent woody plants and charcoal formed during natural wild fires defined as equidimensional (equant) and lath-shaped opaque particles

4.1.1 Amorphous Organic Matter (AOM)

The study and description of the palynofacies constituents of the amorphous organic matter (AOM) prototype of the Nigerian sedimentary basins have attained a series of particulate organic matter detailed account and origin interpretation (Okeke 2024; Okeke et al. 2023a, b, 2017; Okeke and Umeji 2018a, b). The quantitative and qualitative occurrence of amorphous organic matter kerogen elements from the Enugu Formation and Danian sections in this research is typified as fluffy, yellow-amber and dark brown to brown structureless dense substances of dissimilar size ranges, shape along with taphonomy grade. It has been substantiated that the amorphous organic matter kerogen particle is a productive result of bacteria microorganism alteration of gymnospermous and angiospermous land-derived plants, most common in terrestrial, coastal and brackish water environment (Okeke et al. 2023a, b; Okeke and Umeji 2018a, b). However, biological activities of other microbiological organisms of fungi, protozoa and others within the food chain or food web natural ecosystem can relatively breakdown existing plant produce for AOM generation and active oxidizing and poor taphonomic conditions. The origin of AOM form of the studied formations and all marine organic particles formed near the surface of the ocean, within the photic zone, via photosynthesis of microgreen algae is chemically and oxygen-wise broken down or completes the food chain of zooplanktonic organisms. The breakdown of palynofacies elements in marine and non-marine water realm deposited sediments is triggered by microbiological organisms of bacteria, fungi, protozoa and other highly effective oxidizing conditions. Anoxic/stagnant water status is typified by an oxygen deficient content level of < 0.5 ml/l in all water space (Bjorlykke 2010; Knut 2015) for excellent controlled non-degradation and preservation of phytoclasts in sedimentary rocks. Many works (Batten 1983; Duncan and Hamilton 1988; Stemmerik et al.

1990) authenticated the high abundance of amorphous organic matter (AOM) palynofacies material in non-marine sedimentary depositional sequences deposited in anoxia-dysoxia settings.

The shape, characteristic and abundances of the AOM kerogen-type material of the Enugu Formation are controlled by brown to dark-brown medium- and large-sized (15 to > 100 μm) forms of terrestrial origin with slight leading marine AOM particle origin (palynofacies particles 7 and 12 in Fig. 4.1, Table 4.1). The visual microscopic physical characteristics, shape, color and abundances of amorphous organic matter within the Danian sections of the Nsukka Formation are relatively similar to those of the older Enugu Formation. However, the size differs by 25 μm to > 100 μm in the Danian strata.

The sizes, colour and frequent occurrence of the particulate organic matter materials of the Enugu Formation of the Anambra Basin and Danian lithostratigraphic units (Fig. 4.2) signify a major terrestrial source with little marine origin. These details support the AOM kerogen particles of sedimentary rocks reported as products of aquatic and terrestrial origins (Masran and Pocock 1981; Venkatachala 1981; Batten 1996 and Oboh-Ikuenobe et al. 1997). These origin standards of amorphous organic matter are substantiated with fluorescence microscopy and colour changes. Deeper brown colour amorphous organic matter of terrestrial origin is inactive in fluorescence microscopy, while AOM of marine origin fluoresces in visual microscopic studies (Masran and Pocock 1981).

4.1.2 Resin

Resin particle palynofacies group of the studied formations are yellowish to amber colored stem tissues of gymnospermous and angiospermous higher vascular plants with variable sizes and shape. The plant's prevalent shape remains dominated by lath shape and equant shape morphotypes similar to the classification system shape of opaque particulate organic matter. The 10 to > 100 μm size ranges and relative high abundance of resin constituents signify a close resemblance to land-derived plant origin. The catagenic transformation of resin is substantiated by the collection of light yellow to very dark yellow, dark brown and semi opaque colours of the resin particles which signifies gradual diagenesis into opaque particles (Fig. 4.1). However, the structured particulate organic matter alteration into resin constituents is reminiscent of gradual decay of the plant cells and peripheral parts of structured particulate organic matter of land-derived plants. This alteration pattern was observed in visual microscopic studies from gradual degradation grade of structured phytoclast stem tissues to relatively degraded and degraded phytoclast or unstructured phytoclast to resin particle with gradual increase in yellowish to amber coloration morphodynamics at each level of structured particulate organic matter to resin metamorphosis (Fig. 4.1).

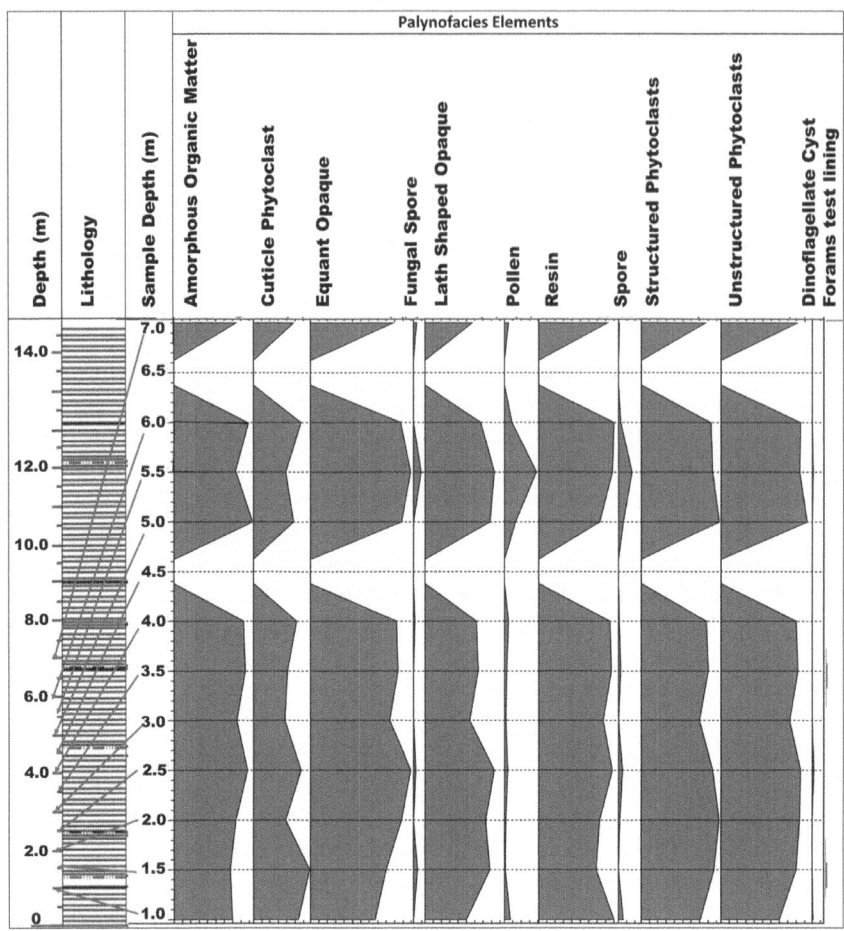

Fig. 4.2 Abundance frequency distribution charts of (%) particulate organic matter constituents from the representative section of the Enugu Formation, at Neke Isi-Uzo vicinity of the Anambra Basin, southeastern Nigeria

4.1.3 Structured (Wood) Phytoclasts

Structured phytoclasts are woody remains of gymnospermous and angiospermous terrestrially derived well-preserved structured and structureless parts of plants. The degree of preservation and abundance of the palynofacies phytoclast is authenticated with large sizes and shades of yellow–brown, brown to dark-brown colours (Fig. 4.1). The palynofacies component display well-preserved conducting tissues and are of various shapes, lath-shaped with occasional equidimensional types with cellular structures and relative dimensional smaller sizes of 10 μm to > 100 μm. Structured phytoclasts are the results of good quality taphonomy standard of the marine

4.1 Origin and Palynofacies Constituents Types of the Enugu Formation ...

processes inherent in the deposition of the Enugu Formation and Danian sediments of the study where terrestrial wood remains from visual microscopic studies are well-preserved with visible conducting tissues and relative large sizes.

4.1.4 Cuticle Phytoclast

Cuticle palynofacies debris analyzed in the Enugu Formation and the Danian sediments are structured flat, platy, thin and well-preserved leaves of angiospermous and gymnospermous vascular plant with various grades of preservation ranging from well-preserved to poorly preserved and relatively degraded with various size ranges of 10 μm to > 100 μm (Figs. 4.1 and 4.3). Brown and dark brown to relatively black colours with various sizes of cellular thickening were documented during the visual kerogen microscopic study of the cuticle phytoclasts. Cuticle phytoclasts with well-preserved epidermal cells exhibit well-defined leaf boundaries with clear upper and lower epidermis. The poorly preserved cuticle phytoclast exhibit pale epidermal outlines diagnostic of little thermal alteration and diagenesis of the encountered cuticle elements that are poorly preserved. The alteration of the cuticle is frequently channeled into resin particles grade via series of epidermal cells degradation and systematic gradual yellowish to amber coloration system.

4.1.5 Unstructured (Degraded) Phytoclasts

The unstructured phytoclasts of the study portray different grades of degradation from partly degraded with little shades of irregular shaped cellular structures and well degraded without any visible structural pattern of the plant remains (Figs. 4.1 and 4.3). Some unstructured forms are partly structured without any visible cell structure. The palynofacies constituent exhibits variable color ranges from brown to dark brown.

4.1.6 Opaque Debris

Opaque particulate organic matter constituents are classified as equidimensional (equant) opaque debris and lath-shaped opaque element and opaque forms without any visible defined shape or structure pattern regarded as opaque particle by Okeke and umeji (2018a, b). These palynodebris are strongly visible in visual microscopic studies. The lath-shaped opaque palynodebris are needle and blade-shaped whereas the equant opaque particles are dominated by angularly shaped forms with relatively continuous angle ± 90 morphotypes with many defined shapes (Figs. 4.1 and 4.3). Most opaque palynofacies classified as equant opaque particles, with irregularly

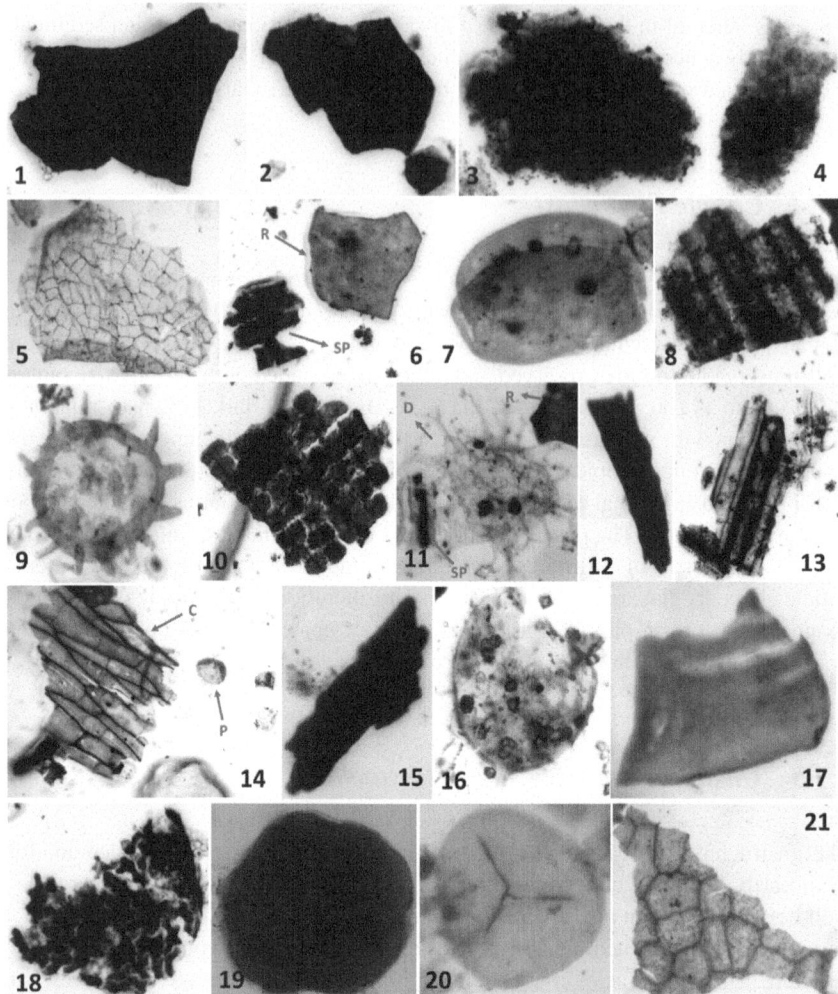

Fig. 4.3 Photomicrographs of palynofacies constituents of the Danian section; *D* Dinoflagellate cyst, *SP* structured phytoclast, *C* cuticle, *P* pollen, *R* Resin. 1. Equant opaque debris 2. Equant opaque debris 3. Amorphous organic matter 4. Amorphous organic matter 5. Cuticle 6. Particulate organic matter components 7. *Longapertites vanendenburgi* Germeraad Hopping and Mulle, 1968 8. Well-structured phytoclasts 9. *Spinizonocolpites echinatus* Muller 1968 10. Degraded black brown phytoclast 11. Palynofacies elements 12. Lath shaped opaque particle 13. Well-structured phytoclasts 14. Particulate organic matter elements 15. Lath shaped opaque particle 16. Dinoflagellate cyst cf. *Damassadinium californicum* (Drugg 1967) Fensome et al. 1993 17. Resin 18. Degraded black brown phytoclast 19. Resin particles 20. *Leiotriletes maxoides maxoides* (Krutzsch) Takahashi and Jux 1989 and 21. Cuticle

4.1 Origin and Palynofacies Constituents Types of the Enugu Formation ...

shaped circular diameter debris reminiscent of opaque particulate organic matter products of charcoal remains from bush burning. However, a large number of the analyzed debris are those of diagenetic thermal alteration (Figs. 4.1 and 4.2) without exhibition of any defined shape (Figs. 4.1, 4.2, 4.3, and 4.4). This was called opaque particle by Okeke and Umeji (2018a, b).

In light of this, opaque debris was considered as remains of charcoal brought about by natural wildfires (Tyson 1993). Opaque particle is regarded as products of oxidation of translucent terrestrial plants material formed during prolonged hydrodynamic transport system of the water realm and post-depositional alteration of particulate organic matter phytoclast of terrestrial origin (Okeke and Umeji 2018a, b). The abundance and relative sizes of the opaque debris from the Enugu Formation and Danian sediments indicate that they are of plant origin (palynofacies elements 3 to 4 in Fig. 4.2). Most of the opaque kerogen in the strata exhibit sole attribute of natural

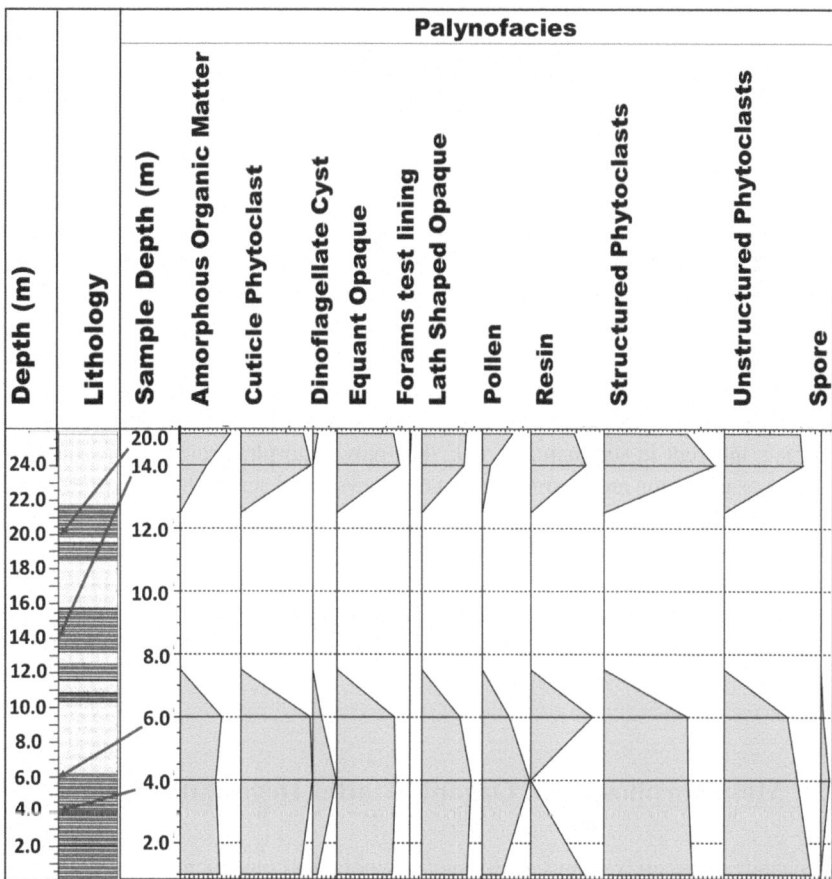

Fig. 4.4 Abundance frequency distribution viewgraph of palynofacies particles from the representative section of the Danian units of the study at Ikpankwu Quarry, southeastern Nigeria

sedimentological diagenesis whereas some display the characteristics of modern charcoal signature as products of terrestrial natural wildfire. The physical physiognomy of the opaque palynodebris of the Enugu and Danian strata demonstrates some attributes of recycled fragments and progressive catagenic transformation inherent in unique black and typical black with yellow–brown to brown circumferential edges, weathered and recycled coal fragments or charcoal from natural forest fires reminiscent of palynofacies particles 7 and 12 in Fig. 4.1. and palynofacies elements 3 to 4 in Fig. 4.3.

4.1.7 Palynomorph Microfossils

Palynomorph was described by previous works as microfossils with very high acid (HCl and HF) resistant sporopollenin, chitin or "pseudochitin" organic molecule recovered from sediment after chemical maceration (Traverse 2007). In the contest of particulate organic matter analysis, palynomorphs are group of palynological constituents of marine and non-marine (terrestrial) origin which is defined in practical and microscopic visual analysis as microorganism constituents encountered in palynological microscopic studies after chemical maceration techniques of using HF for removal of silicates and HCl for removal of carbonates from sedimentary rock during laboratory analysis as applied in this research (Fig 1.4). It can also be described as different microorganism elements encountered in the microscopic slide/studies after standard analytical techniques of nonacid or acid model. This proves and authenticates the two standard accepted palynology maceration techniques and the residual imprints as palynomorphs and palynofacies phytoclasts/particulate organic matter debris. These are based on various analytical objectives and sandstone digestion processes of oxidation and non-oxidation methods of plant materials inherent in sedimentary rock. The palynomorph species groups in palynofacies examination are marine dinoflagellate cysts and acritarch, terrestrial pollen and spore microflora (including algal and fungus spores), foraminifera test linings along with any other microorganism prototype that would be encountered during palynology microscopic studies. The color ranges of the palynomorph prototypes encountered in the cause of the analysis of the Enugu Formation and Danian unit sediments are primarily brown and dark brown and relatively yellow in colour with an average of 30 μm to > 100 μm.

4.2 Metamorphosis and Organic Matter Degradation

The change in the study's physical form of palynofacies constituents necessitated the appraisal of the particulate organic matter diagenesis imprints of the Enugu Formation and Danian section palynofacies components. The degree of metamorphosis and alteration of particulate organic matter concept can be measured by the hydrodynamic, thermal maturation, organic matter saltation and depositional setting of the palynodebris. Thermal alteration of the plant debris with an increase in temperature alters the original parent plant color from dull-colored material to relatively shiny debris and dark brown-black colour with an increase in burial depth and temperature. The thermal alteration index can be quantified by the degree of vitrinite reflectance index or light reflected from the plant material in visual microscopic examination processes. The vitrinite reflectivity of palynodebris upsurges with higher temperatures and rate of maturity. The degree of organic thermal alteration index vitrinance reflectance model was illustrated in kerogen maturation synthesis analysis (Pearson 1984; Travers 2007; Okeke and Umeji 2018a, b; Okeke 2017, 2023).

In petroleum geochemistry analysis, vitrinite reflectivity of 1.2 reflects high oil generation of the source rock indicative of middle level oriented "oil window" whereas a 0.7–0.8 and lesser vitrinite reflectance values are inherent in source rocks with low heat or temperature rate designated as immature source rocks in Rock Eval pyrolysis and kerogen maturation studies (Bjorlykke 2010; Pearson 1984; Hayes et al. 1983. Traverse 2007). However, a vitrinite reflectivity of nearly 2.0 and above suggests a source rock archetype with maximum exhaustive oil generation and systematic capacity to generate only gas in the petroleum system. The vitrinite reflectivity, thermal maturation and organic matter metamorphosis are apparent in the degrees of black with yellow–brown to dark brown circumferential edges color of the opaque debris. The organic matter degradation status of the studied Anambra Basin formations ranges from relatively small to medium and large sizes of the debris because the sizes, shapes and quality of palynofacies elements are the products of degradation. These degradation characteristics of plant debris are substantiated by the hydrodynamics and saltation systems of plants inherent in wave and current generated energy of the marine realm. The degree of degradation in particulate organic matter is also quantified by the numerical value and frequency of particulate organism sizes inherent in the palynomaceral scheme of Whitaker (1982, 1984) for basic hydrodynamic and saltation mechanics cum depositional environment criteria for the grade of palynofacies alteration within the fluvial and marine water realm.

4.3 Paleoenvironment

The detailed palynofacies, lithofacies and sedimentary structures attributes synthesis investigation of the Enugu Formation of the Anambra Basin and Danian lithostratigraphic units established the palynofacies hydrodynamics, sandstone saltation, water depth and depositional environment conditions of the basin during the Late Campanian to Early Maastrichtian and Danian sedimentary processes of the formations. The diverse sedimentary analytical parameters inherent in lithofacies limits of grain size textural quality, sedimentary structure and palynofacies constituents characteristic of the outcropping sediments engineered the palynofacies depositional environment and palynofacies hydrodynamics system for high resolution paleoenvironment dynamics reconstruction. The paleoenvironment system of the study buttressed the abundance of well-preserved and degraded land-derived palynofacies microflora and marine palynofacies debris components. This palynofacies model perfectly enhanced the kerogen maturations system, hydrocarbon generation potential, structural style dynamics and depositional environment design triggered during basin evolution with a series of progradation and retrogradation packages of sedimentary sequences during the Upper Cretaceous and Early Paleocene.

4.3.1 Palynofacies Depositional Environment of the Enugu Formation

The paleoenvironment reconstruction of the Enugu Formation sedimentary rocks of the Anambra Basin unravels the systematic dynamic attributes of particulate organic matter elements, depositional facies and sedimentary structure style of the formation (Table 4.2). These dynamic natural biologic and physical geology events triggered the high resolution detailed palynofacies and sedimentary rock sandstone grains hydrodynamics synthesis, water depth relationship and depositional environment of the formation. These geology events illustrate the Late Campanian to Early Maastrichtian stratigraphic age buttressed in time and space geology concept of the Enugu Formation. Source rock, depositional environment and sequence stratigraphic events of oil-producing palynofacies group were reported (Okeke et al. 2021; Okeke 2023; Tyson 1995; Travers 2007; Batten and Stead 2005; Akande et al. 2007; Batten 1996) as the hydrodynamic effects of sedimentary rock and organic matter system.

Table 4.2 Synoptic description of some crucial palynofacies terminology in the appraisal of provenance and paleoenvironment settings

Terminology	Definition
Palynofacies	The quantitative and qualitative palynological study of the total particulate organic matter present in a sedimentary rock
Palynomorph	Palynomorph is a high chemical (HCl and HF) resistant microfossil with sporopollenin, chitin or "pseudochitin" organic molecule recovered from sediment after the chemical maceration technique
Phytoclast	Angiospermous and gymnospermous terrestrial plants structured and relatively unstructured dispersed clay to fine-grained sandstone sized palynofacies particles
Palynofacies palaeoenvironment	It is the reconstruction of the palaeoenvironment of deposition of sedimentary rocks based on the plants hydrodynamic and fluvial mechanics of the water realm
Palynofacies provenance	Palynofacies provenance is concerned with the geographic origin, fluvial and marine hydrodynamic processes of plant remains along with the deposition of sand-sized sediments

4.3 Paleoenvironment

The paleoenvironment and hydrodynamics system of the palynofacies constituents of the Enugu Formation are episodic events that are triggered by the relative fluctuations of key depositional environment processes inherent in the particulate organic matter sizes and abundance, championed by paleoenvironment and paleoclimate oscillations during sediment deposition.

The lithofacies of the studied key structural style oriented outcrop sequences of the Enugu Formation (Figs. 3.1, 3.2, 3.3 and 3.4) of the Amagu road cut outcrop, Aguabo outcrop by the Aguabo pedestrian bridge road cut, Ozalla/Four-Corner road cut and Neke Isi-Uzo area are characterized by medium- to fine-grained sandstones and shale, siltstones, mudstone facies deposited in swaying depositional settings of the outer neritic, brackish and fresh water settings deposited within a more pronounced inner neritic coastal marine environment (Fig. 4.5).

The depositional environment pattern of the outcropping sections of the formation substantiated the palynofacies hydrodynamic and lithofacies events and the dominant basal carbonaceous mudstone lithofacies and heterolithic units with intermittent beds of siltstone and medium- to fine- grained sandstone facies (Umeji 2000, 2005; Okeke et al. 2023a, b). The shale and rhythmically mudstone successions are regularly interbedded with relatively ferruginized to white and grey siltstones along with ironstone units and bands that grade into 2 m thick heterolithic sedimentary rock succession in some places. The heterolithic section of the formation is dominant in the type area around Enugu suggesting a generic shallowing upward system of the lithofacies into tidally influenced shoreface and marshes depositional environments but with intermittent water-depth variations.

The shale facies of the Enugu Formation are typified by massive high occurrence of AOM, and few marine microflora, structured and unstructured phytoclasts, along with pollen and spores microflora indicative of an inner neritic marine paleoenvironment with infrequent outer neritic influences (Figs. 4.1 and 4.5). Palynomorphs, palynofacies (Okeke et al. 2023a, b), micropalaeontology and other macropalaeontological occurrences of foraminifera (Reyment 1965; Agagu et al. 1985; Umeji 2000) along with ammonite, gastropod, bivalve and fish teeth (Zarboski 1982, 1983) are evidence of mixed water-depth models inherent in the Enugu Formation. The occurrence and abundance of terrestrially derived particulate organic matter (> 90%) in the formation relates the palaeoecology and rapid evolution perspective of angiospermous and gymnospermous higher vascular plants in the Late Campanian to Early Maastrichtian stratigraphic period of the formation to a favorable quality depositional environment and remarkable quantity of plant remains (Fig. 4.2). The siltstone sedimentary successions of the study are products of inner neritic or tidally influenced channels while the heterolith lithofacies units are of intertidal or subtidal coastline depositional settings. This buttressed the prevalence of palynofacies, palynomorphs, and other micro and macro palaeontology evidence reported in this research.

The diagnostic sedimentary rock grain size hydrodynamic traces of deltaic progradation depositional trend for regression in an active delta growth were indicated as the depositional setting of the mudrock and medium- to fine-grained sandstone lithologic successions of the formation (Okeke et al. 2023a, b; Ojoh, 1990; Ojo et al. 2003; Ojo et al. 2009). This suggests that the full marine prodelta subfacies basal

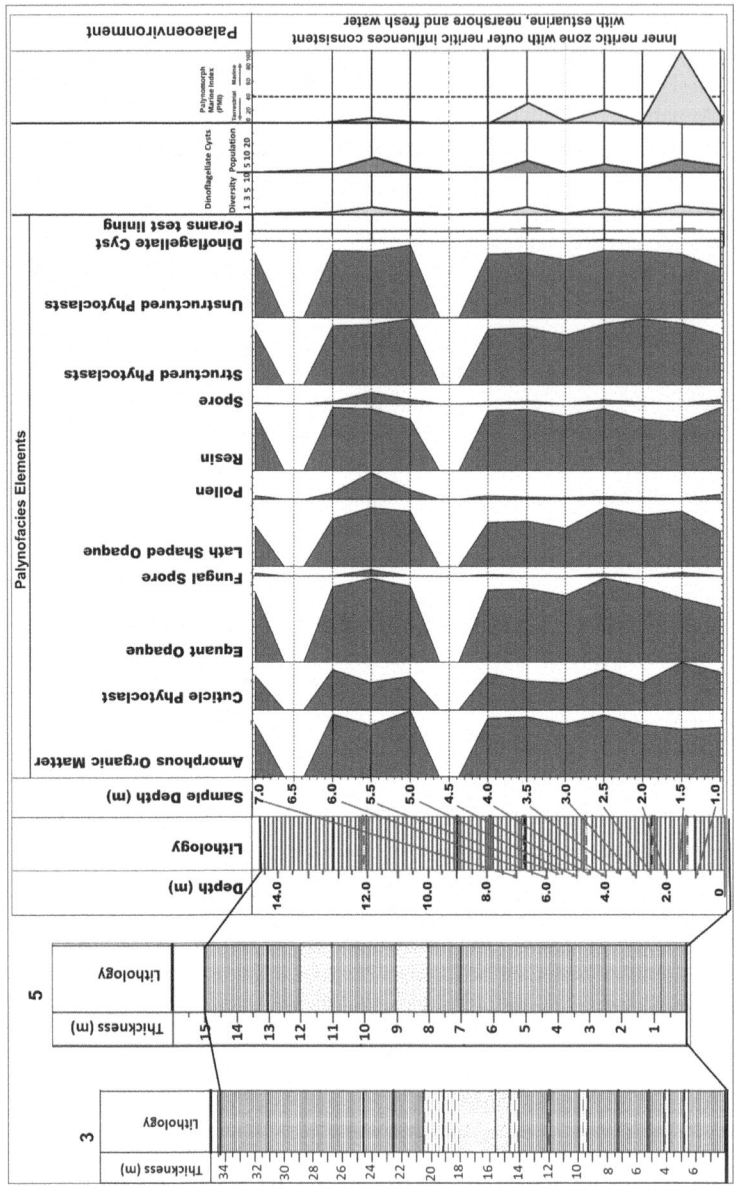

Fig. 4.5 Summary of the particulate organic matter chart of the relative abundance of palynofacies elements distribution and depositional environment of the Enugu Formation of the Anambra Basin at the Neke Isi-Uzo outcroping section, southeastern Nigeria, showing pollen and spore frequencies, palynofacies index, dinoflagellate cyst indices and abundances, lithostratigraphic surfaces, and Palynomorph Marine Index (PMI) significance

4.3 Paleoenvironment

mudrock sediments shallows upwards into tidally influenced shoreface and marshes setting. This type of palaeoenvironmental setting was illustrated in the subcropping Niger Delta basin where the highest TOC and Hydrogen Index (HI) values in non-marine swamp, marsh and floodplain suggests proximity of the depositional space to organic source (Bustin 1988). The shallow marine settings identified in the study area are confirmed by the abundance of land-derived palynofacies groupings (Figs. 4.1 and 4.2). The sideritic carbonaceous shale facies component of an open marine environment (Reijers 1996) combined with dinoflagellates and AOM (Umeji 2000; Okeke et al. 2023a, b) support the outer neritic quality of the Enugu Formation whereas benthonic foraminifera genera of *Gavelinella*, *Planulina* and *Cibicides* were the key shelf environment dwellers (Ojo et al. 2003; Kolawole 2004; Peters 1983) within an inner neritic setting. The high grade of the taphonomy of terrestrially derived organic matter (> 90%) and lowest occurrence of marine microflora species are characteristic of the Enugu Formation settings (Fig. 4.2). The microenvironments of the inner neritic settings of the study are substantiated by the non-marine swamp, marsh and floodplain environments which denotes distinct proximity of the deposited strata to an organic source. These paleoenvironmental settings are ranked at the top of the productive vegetation zones' chart, creating diverse quality taphonomy organic matter observed in the Enugu Formation sediments.

The paleoenvironment analysis of the combined palynofacies and lithofacies systems of the Zubair Formation in Iraq (Al-Ameri and Batten 1997; Ali and Nasser 1989) confirmed similar palynofacies hydrodynamic distributions with the Enugu Formation swamp and marsh depositional environments, oscillating from delta top, delta front, prodelta and open marine conditions within the region. The low frequency of dinoflagellate cysts and other marine taxa with high diversity and abundance of land-derived microflora in the Enugu Formation suggests the brackish water settings which is equivalent to a shallow marine and deep marine interactive interface similar to shallow marine influenced deep marine settings definitive of coastal environment oscillation of the water realm (Fig. 4.5).

4.3.2 Palynofacies Depositional Environment of the Danian Lithostratigraphic Units

The Danian lithostratigraphic unit sedimentary successions of marine and non-marine lithofacies paleoenvironment reconstruction model is ideally suited to interpret combined palynofacies and depositional facies analysis aimed at palaeoenvironmental interpretation, relative palynofacies provenance and hydrodynamic pattern of the formation. The sedimentary rock hydrodynamic system and particulate organic matter quality taphonomy grade of the study reflects the abundance of well-preserved terrestrial palynofacies phytoclast (90%) over marine particulate organic matter (10%), indicative of shallow marine depositional settings with outer neritic influences (Fig. 4.6). The Nsukka Formation depositional cycle contains predominant terrestrial phytoclasts and pollen and spore microflora in the Late Maastrichtian dominant coal facies and Danian predominant mudrock facies forming environment. Prevalence of prograding sequences and transition to early Cenozoic rocks succession was deposited after the retrograding sequences of the late Cretaceous age. This authenticates and confirms the previously reported northerly coaly and southerly shale paleoenvironment (Okeke et al. 2023a, b; Okeke 2024; Umeji and Edet 2008; Umeji 2000, 2005; Umeji and Nwajide 2007) dictates of the formation that is aligned to the depositional and chronostratigraphic timeframe of the Nsukka Formation. The lower and upper deltaic plains fluctuating from tidal flat, lagoon, tidal bar, raised bog and reed swamp northerly paleoenvironment settings to nearshore open marine southerly linked depositional environment depicts the oscillating paleoenvironment systems of the formation (Umeji and Edet 2008; Umeji 2000, 2005; Umeji and Nwajide 2007). The paleoenvironment interpretation of the Zubair Formation Iraq, projected a synthesized palynofacies and lithofacies prototype that documented similar palynofacies hydrodynamic distributions on swamp and marsh depositional settings oscillating from delta top, delta front, prodelta and open marine ecologic conditions (Al-Ameri and Batten 1997; Ali and Nasser 1989). The fluvial, estuarine and deltaic environments resultant products of tides, current and wave effects of the marine realm are the key products of the lithofacies units of the Nsukka Formation with consistent vast sedimentary rock succession dissimilarity and rapid lithological change substantiated in the series of studied coal, medium- and fine-grained sandstone, heteroliths, parallel laminated siltstone and shale with mudrock facies and crossbedded sandstone in the outcrop units at Ihube, Okigwe and Ikpankwu vicinity (Fig. 3.4).

The palynofacies quantitative dimensions of the mudrock lithofacies successions indicated a relatively high frequency of structured terrestrial particles and degraded plant remains in shale units (Tripathi and Divya 2012). This concurs with the quality and abundance of the particulate organic matter elements of the shale lithologies of the Ikpankwu, and Ihube vicinity outcropping units typified in the high abundance of medium- to large- sized structured and unstructured phytoclasts and pollen and spores of terrestrial origin over marine palynomorph and relatively medium- to small-sized AOM of the study (Figs. 4.3 and 4.6). The high energy settings of the Danian water realm triggered the plane parallel lamination, heteroliths of siltstone and shale and erosional base with rip-up clasts lithofacies deposits which also depicts a systematic sporadic flooding event in the upper deltaic plain depositional space. Extensive production of particulate organic matter is denoted in these settings where angiospermous plants and trees are frequently preserved by sinking into mud or bog without oxygen. Umeji (2005) denoted the prevalence of flaser and lenticular bedding sedimentary structures of the Nsukka Formation due to scarcity of sand and clay respectively within the flood plain or vicissitude in the energy of the transportation medium. The mudrock-dominated heterolith facies suggest changing current depositional systems (Reineck and Singh 1980). However, the absence of dinocyst marine microflora taxa in the variety of mudrock facies of the Nsukka Formation refutes tidal influences (Okeke et al. 2023a, b). The incidence of dinoflagellate cysts and AOM of marine origin along with the dark to gray shale facies depicts distal anoxic to suboxic environmental settings source of the palynofacies constituents of

4.3 Paleoenvironment

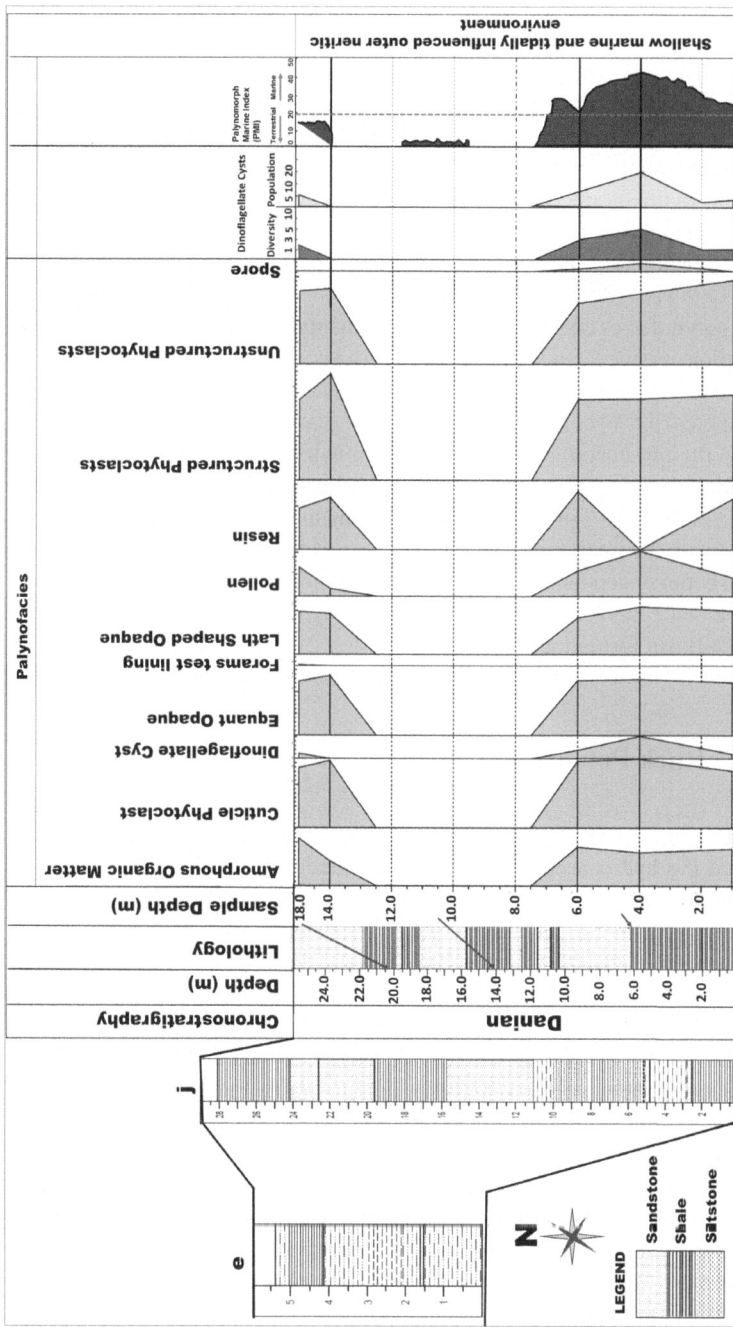

Fig. 4.6 Stratigraphic summary of the palaeoenvironment and palynofacies distribution viewgraph of the relative abundance of particulate organic matter elements at the Ikpankwu outcrop in the southern flank of the Danian lithostratigraphic section, southeastern Nigeria, showing pollen and spore frequencies, palynofacies index, dinocyst indices and abundances, lithostratigraphic surfaces, Palynomorph Marine Index (PMI) significance and depositional environment

the study inherent in open marine lagoon or brackish settings (Fig. 3.4). However, few AOMs encountered in the study depicts aquatic origin showing some marine influences with slightly fluffy to dark and yellowish brown to black brown colours (Figs. 4.3 and 4.6). The AOM of the Danian units is marked by the prevalence of AOM particles of marine origin typified by fluffy, yellow-amber and brown structureless dense humic gel-like materials of dissimilar size ranges with well-defined degrees of preservation and shape.

The medium- to fine-grained sandstone with kaolinitic clay and a series of rip-up clast quantify the intensity of oxidation, recycling, and forest fires of terrestrial plants which validates the high occurrence of opaque phytoclast indicative of the proximal shelf settings (Gorin and Steffen 1991; Jaramillo and Oboh-Ikuenobe 1999). The nonoccurrence of palynofacies constituents in this sedimentary rock facies of kaolin, and coarse- to medium-grained sandstone is because oxygen diffuses to about 5–20 cm depth below the seabed within coarse- to medium-grained sandstone sediments. Pteridophyte fern spores (0.5%) of hygrophilous plants of acanthus origin from freshwater algae illustrate the presence of local or seasonal humid condition during sediment deposition (Schrank and Mahmoud 2000). The synthesized lithofacies and kerogen assessment of the Nsukka Formation Danian sedimentary successions engineered a systematic, detailed comprehension of this aspect of integrated sedimentary rock processes for a better understanding of the paleoenvironments and changes in relative sea level as the delta evolved to a significant major hydrocarbon province inherent in the Anambra Basin formations.

4.4 Source Rock Potential

The palynofacies model of the Enugu Formation and the Danian horizon of the study enhanced the hydrocarbon exploration and exploitation exposé, to improve the understanding and optimistic impact of the high-resolution palynofacies analysis replica of Anambra Basin, for potential hydrocarbon generation prospects. Petroleum generation prospective measures can be linked to the development of oil fields for liquid oil and gas probing campaigns in the southeastern region of Nigeria. The kerogen types and organic thermal maturity indices of the Enugu Formation and the Danian strata are shown in this work whereas the palynofacies abundance distribution statistical frequency charts of the Enugu Formation around the Neke Isi-Uzo area and the Danian section in the vicinity of Ikpankwu is displayed in Figs. 4.2 and 4.4 as an interpretation standard outcrop, statistical generated viewgraph model of this research. Organic geochemistry, the study of organic matter in sediments and its dynamic hydrocarbons transformation processes is expatiated for an advanced exposé of the palynofacies quality and quantity, organic thermal maturation and source rock perspective synthesis of the studied formations. The organic matter remains of land and marine origin is broken down via direct oxidation from $CH_2O + O_2 \rightarrow CO_2 + H_2O$ and microbiological organic food chain processes. The prevalence of oxygen obliterates organic matter rapidly on land and in the sea with availability of oxygen.

4.4 Source Rock Potential

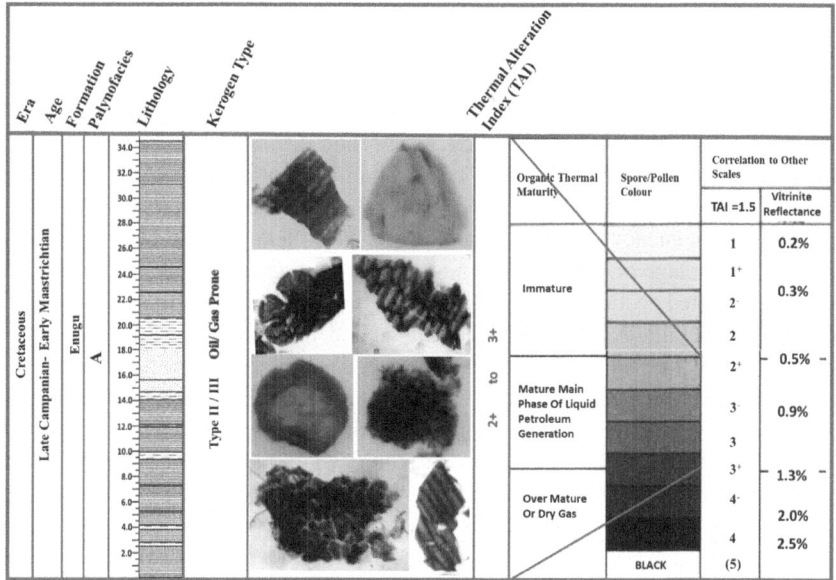

Fig. 4.7 Palynofacies, kerogen types and organic thermal maturation of the Enugu Formation, Anambra Basin

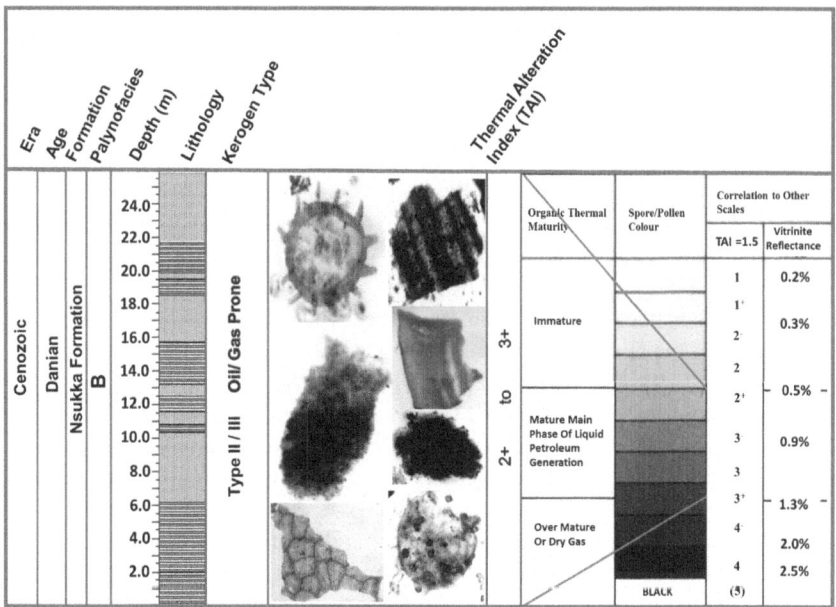

Fig. 4.8 Particulate organic matter, kerogen types and organic thermal maturation of the Danian unit of the study from the Ikpankwu vicinity, southeastern Nigeria

The Enugu Formation palynofacies components shows that the particulate organic matter is indicative of oil/gas prone Kerogen Type II/III which is a product of quality and quantity of high abundance of large- to medium-sized well-preserved structured phytoclasts (14.5%), lath-shaped and equant opaque debris (30%), well-preserved medium- and large-sized brown cuticle (8.6%) and degraded wood (Figs. 4.1, 4.3 and 4.7). They occur in association with pollen and spore (0.6%), dinoflagellate cysts (< 1%) and relative frequency of amorphous organic matter (14%). These palynofacies frequency distribution prototypes from Neke Isi-Uzo outcrop section fall within the range of particulate organic matter abundance attributes encountered in Amagu faulted section, Aguabo outcrop by the Aguabo pedestrian bridge road cut, and the Ozalla/Four-Corner road cut faulted horizon of the Enugu Formation (Fig. 4.3). The high palynodebris occurrences indices of the study portray a significant high abundance of terrestrial palynomorph and structured phytoclasts of land origin over marine palynomorph and amorphous organic matter constituent which illustrates rich availability and proximity of particulate organic matter of terrestrial origin with little and relatively distant AOM and dinocysts of marine origin (Fig. 4.3). These authenticates the quality of palynofacies imprints and the significant effective primary productivity, biochemical degradation and depositional environment dynamics of sedimentary rock in association with the hydrodynamics pattern of organic matter in sediments.

The palynofacies constituents of the prototype Danian unit of the Nsukka Formation at Ikpankwu area are made up of quality and quantity of fluffy, dark brown and relatively brown AOM (>8%), dinoflagellate cyst species of *Kenleyia lophophora, Spiniferites ramosus, Kenleyia leptocerata*, trochospiral forams test lining along with other marine biotas, large- to medium-sized well-preserved brown to dark brown structured phytoclasts (20%), cuticle (15%), lath-shaped and equant opaque debris (20%), and medium- and small-sized degraded wood remains (Fig. 4.3). These kerogen frequency characteristics of the study is culminated in the high abundance of structured phytoclasts and pollen and spores microflora of land origin over marine species and amorphous organic matter (10%). The statistical palynofacies frequency measures illustrates proximity to terrestrial origin's existing plant source vegetation ecosystem rather than marine origin (Fig. 4.4). The palynofacies components of Danian units of the Nsukka Formation show that the organic matter elements are of Kerogen Type II/III which is Gas/Oil Prone (Figs. 4.4 and 4.8). The kerogen imprints of the study depict Kerogen Type III (gas prone) substantiated by little amount of AOM terrestrial and marine origin with high abundance of terrestrial phytoclasts and palynomorphs. The Oil-Prone Kerogen Type II palynofacies group are substantiated relative to marine AOM and dinoflagellate species and marine other biota. The marine and non-marine particulate organic matter disequilibrium is substantiated in the high occurrence and diversity of pollen and spores with well-preserved large phytoclast palynofacies constituents and opaque debris demonstrative of deposition close to parent source vegetation, expected of the good quality and size ranges of the palynodebris. Generally, the palynodebris inclination and dimensions of the Danian sediments has a punctuated uneven equilibrium land derived phytoclasts, pollen and spores along with amorphous organic matter and other marine biotas (Fig. 4.4).

4.4 Source Rock Potential

The palynofacies components of the Enugu Formation and the Danian stage proves that the particulate organic matter quality and quantity of the studied part of the Anambra Basin portrays an admixture of kerogen Type II/III which is oil/gas prone (Figs. 4.7 and 4.8). The statistical plot of HI versus Tmax of Ojo et al. (2009) reappraises the Type III kerogen group while the soluble organic matter component ranges of 578–1931 ppm suggests fair to good source rock prospect of the Enugu Formation. The relative abundance and quality of AOM, dinocyst and foraminifera test linings of marine origin indicates good source rock suggestive of good oil shale potential. Maximum flooding surfaces inherent in condensed sections of sequence stratigraphy packages are the best source rock (Tyson 1995; Batten 1999) while TOC constituents of mudrock facies, and amorphous organic matter of marine realm algal origin were regarded as an unprecedented good quality source rock (Traverse 2007). However, the quality and quantity of TOC slightly negate high petroleum source potential because carbon material could contain dark woody terrestrial plant materials that exhibits very poor indication for oil (Mehrota 2002). High abundance of the amorphous organic matter of the examined formations, prevalence of dinocysts and an average of about 3 km reported thickness of the Enugu Formation (Grove 1951; Ojo et al. 2009) and a general thickness of about 270 m (Umeji 2000), 400 m (Nwajide 2013) of the Nsukka Formation along with the relatively small- to large-sized well-preserved AOM, marine palynomorph microflora and terrestrial kerogen elements are the qualitative potential indicators of good source rock for Kerogen Type II/III of the studied formations. The particulate organic matter attributes of the younger Mamu and Nsukka formations of the Anambra Basin were indicated as vital prospects pointer of Type II/III kerogen (Okeke 2023). It is important to note that old organic material is systematically buried by the gradual accumulation of the overlying sediments along with the systematic expellant of water during sediment compaction.

In the light of this, marine palynofacies elements of AOM, foraminiferal test lining and dinoflagellate cyst (Figs. 4.1, 4.2, 4.3 and 4.4) are the key products of Kerogen Type II, which is Oil Prone. In contrast, the terrestrial palynomorphs, unstructured and structured phytoclasts, and palynodebris groups are the decisive factors of gas-prone Kerogen Type III. However, allochthonous organic matter of terrestrial origin is indicated as a major source of oil and gas, particularly in an equidimensional marine and terrestrial-prone deltaic and non-marine settings (Thomson 1982).

Figures (4.7 and 4.8) portrays the kerogen types and organic thermal maturity of Enugu Formation and the studied Danian stratigraphic column of the Nsukka Formation. The scales in this pictured chart are correlative to the spore/pollen colour standard chart of Pearson (1984), calibrated to additional organic thermal maturation techniques in Traverse (2007); which is essential in determining the thermal alteration index (TAI) and vitrinite reflectance (%Ro) of the studied formations. The Thermal Alteration Scale (TAS) of Batten (1980) was equally correlated to Pearson's (1984) spore/pollen standard chart for palynomorph colour and its important definition schematics in hydrocarbon maturity index (Figs. 4.7 and 4.8). The relatively abundant brown to dark brown visual palynofacies particles of terrestrial origin from the formations are the factual effects of increased medium to high-temperature events, which are the permissible results of biodegradation of the palynofacies materials and thermal alteration of particulate organic matter.

Kerogen temperature maturation inclination of plant material, absence of palynofacies evolution and extinction dimension schematics rudimented the visible resemblance of kerogen dynamics and systems of the formations in the region. These basic principles between the Cenozoic outcropping succession of the Niger Delta Basin formations (i.e. Imo, Nanka, and Ogwashi formations) and the subcroping Akata and Agbada formations were obvious in geology studies. These quality palynofacies relationships were also indicated in the Upper Cretaceous Anambra Basin formations (Enugu and Nsukka formations). This is because the palynofacies and organic thermal maturation pattern reported indices of these basins exhibit similar standard kerogen quality in rock Eval pyrolysis data and palynofacies data (Okeke and Umeji 2016, 2018a, b; Okeke et al. 2023a, b; Okeke et al. 2021; Ojo et al. 2009; Nwachukwu and Chukwura 1986; Bustin 1988; Claret et al. 1981). These similarity indices of the basins were also noted in similar Rock Eval and palynofacies analysis, kerogen types, quality and quantity of the particulate organic matter, with even paleoenvironment style of the Anambra and Niger Delta basins in the area (Short and Stauble 1967; Bustin 1988; Agagu and Ekweozor 1982; Ojoh 1990; Okeke and Umeji 2016, 2018a, b; Okeke et al. 2023a, b).

In thermal alteration system of source rock, it is important to note that gas is formed later than oil which also generates and migrates later than oil. At this juncture, changes in the colour of organic matter uphold the degrees of thermal alteration of palynodebris pollen and spores components in sedimentary rock successions. Spore/pollen colouration index of the Enugu Formation in the Neke Isi-Uzo and Danian Nsukka Formation at the Ikpankwu vicinity showed qualitative grade of the Thermal Alteration Index (TAI) of 2 − to 3 +, Vitrinite Reflectance Index (%Ro) of 0.3–1.3% and organic thermal maturation of mudrock facies suggestive of low- to intermediate-thermal maturation condition. This Thermal Alteration Index (TAI) and Vitrinite Reflectance Index measures correenspond to the spore/pollen colouration of relatively light brown to dark brown colour. The colour range signify "mature main phase of liquid oil generation" to the production of wet gas and transition to dry gas phase hydrocarbon generation schematics (Figs. 4.7 and 4.8). This is synonymous with the standard thermal maturation colour and Thermal Alteration Index (TAI) 2 + to 3 + along with the Vitrinite Reflectance Index (%Ro) 0.5–1.3% (Pearson 1984). The standard %Ro scheme of Hayes et al. (1983) which falls within 50–150 catagenic (Ro: 0.5–2.0) is equal to the Vitrinite Reflectance Index (%Ro) range of the formations. Tmax standard of 440–455 °C denoted from the mudrock facies Rock–Eval pyrolysis studies of the Barren Measures Formation in the Kulti (Ku) area portrays an early mature to post mature stage while 432–448 °C Tmax measure from the Sitarampur (Si) areas all in Raniganj Basin suggests early mature stage respectively (Hazra et al. 2015; Varma et al. 2015) similar to the thermal maturity index recorded in the studied formations.

The pollen and spore colour connection and kerogen hydrocarbon generation and expulsion quality model was illustrated in the series of primary palynofacies analysis (Staplin 1977; Hart 1986; Welte and Tissot 1984). The Kerogen Type III mark of the palynofacies elements of the Enugu Formation and the Danian Nsukka Formation is similar to the TOC numerical values of 11.20 and 22.78 wt% and thermally mature (Tmax changes from 461 to 602 °C) thermal alteration index of Barakar Formation and Barren Measures mudrock facies suggestive of condensate/wet-gas and dry-gas windows in thermal maturity arrays reported experiments of Hazra et al. (2020). A maximum uneven absolute thermal alteration or perfect combustion index of Type III land-derived kerogen was illustrated as an out final temperature of 850 °C and lower oxidation greater temperatures of 750 °C with lower sample weights (Hazra et al. 2017; Lafargue et al. 1998; Sykes and Snowdon 2002).

References

Agagu OK, Ekweozor CM (1982) Source rock characteristics of Senonian shales in the Anambra Syncline, Southern Nigeria. J Min Geol 19(2):51–61
Agagu OK, Fayose EA, Petters SW (1985) Stratigraphy and sedimentation in the Senonian Anambra Basin of Eastern Nigeria. J Min Geol 22:25–36
Akande SO, Ogunmoyero IB, Petersen HI, Nytoft HP (2007) Source rock evaluation of coals from the Lower Maastrichtian Mamu Formation, SE Nigeria. J Pet Geol 30(4):303–324
Al-Ameri TK, Batten DJ (1997) Palynomorph and palynofacies indications of age, depositional environments and source potential for hydrocarbons: Lower Cretaceous Zubair Formation, southern Iraq. Cretac Res 18(6):789–797
Ali AJ, Nasser MO (1989) Facies analysis of the Lower Cretaceous oil-bearing Zubair Formation in southern Iraq. Mod Geol 13:225–242
Batten DJ, Stead D (2005) Palynofacies analysis and its stratigraphic application. In: Koutsoukos EAM (ed) Applied stratigraphy. Springer, The Netherlands, pp 203–226
Batten DJ (1980) Use of transmitted light microscopy of sedimentary organic matter for evaluation of hydrocarbon source potential. IV Int Palynol Conf Lucknow (1976–77) 2:589–594
Batten DJ (1983) Identification of amorphous sedimentary organic matter by transmitted light microscopy. Geol Soc 12(1):275–287, London, Special Publications
Batten DJ (1996) Palynofacies and palaeoenvironmental interpretation. Palynol Principles Appl 3:1011–1064
Batten DJ (1999) Small palynomorphs. In: Jones TP, Rowe NP (eds) Fossil plants and spores. Modern Techniques Geological Society, London, pp 15–19 (Geol Soc London Special Publications 12(1):275–287)
Bjorlykke, K (2010) Petroleum geoscience: From sedimentary environments to rock physics. Springer Science & Business Media
Bustin RM (1988) Sedimentology and characteristics of dispersed organic matter in tertiary Niger Delta: origin of source rocks in a deltaic environment. AAPG Bull 72:277–298
Claret J, Jardine S, Robert P (1981) La Diversite Des Roches Meres Petrolieres: Aspects Geologiques Et Implications Economiques a PartirExemples. Bull Cent Rech Explor Prod Elf Aquitane 5:383–341
Combaz A (1964) Les Palynofacies. Revue De Micropaleontolgie 7:205–218
Duncan AD, Hamilton RFM (1988) Palaeolimnology and organic geochemistry of the Middle Devonian in the Orcadian Basin. Geol Soc Lond Spec Publ 40(1):173–201
Gorin GE, Steffen D (1991) Organic facies as a tool for recording eustatic variations in marine fine-grained carbonates—example of the Berriasian stratotype at Berrias (Ardèche, SE France). Palaeogeogr Palaeoclimatol Palaeoecol 85(3–4):303–320
Grove AT (1951) Land use and soil conservation in parts of Onitsha and Owerri provinces
Hart GF (1986) Origin and classification of organic matter in clastic systems. Palynology 10:1–23

Hayes JM, Wedeking KW, Kaplan IR (1983) Precambrian organic geochemistry-preservation of the record

Hazra B, Varma AK, Bandopadhyay AK, Mendhe VA, Singh BD, Saxena VK, Samad SK, Mishra DK (2015) Petrographic insights of organic matter conversion of Raniganj basin shales, India. Int J Coal Geol 150–151:193–209

Hazra B, Dutta S, Kumar S (2017) TOC calculation of organicmatter rich sediments using Rock-Eval pyrolysis: critical consideration and insights. Int J Coal Geol 169:106–115

Hazra B, Wood DA, Singh PK, Singh AK, Kumar OP, Raghuvanshi G, Singh DP, Chakraborty P, Rao PS, Mahanta K, Sahu G (2020) Source rock properties and pore structural framework of the gas prone Lower Permian shales in the Jharia basin. India Arab J Geosci 13(13):1–18

Ibrahim MIA, Abul Ela NM, Kholeif SE (1997) Paleoecology, Palynofacies, Thermal Maturation and Hydrocarbon Source-Rock Potential of the Jurassic-Lower Cretaceous Sequence in The Subsurface of the North-Eastern Desert, Egypt. Qatar University, Sci J 17(1):153–172

Jaramillo CA, Oboh-Ikuenobe FE (1999) Sequence stratigraphic interpretations from palynofacies, dinocyst and lithological data of Upper Eocene-Lower Oligocene strata in southern Mississippi and Alabama, US Gulf Coast. Palaeogeogr Palaeoclimatol Palaeoecol 145(4):259–302

Knut B (2015) Petroleum geoscience: from sedimentary environments to rock physics

Kolawole AU (2004) Source rock characteristics and biostratigraphy of the Campanian—Maastrichtian Enugu and Mamu formations, Enugu, Anambra Basin. Unpublished M.Sc. Thesis. University of Ilorin, Ilorin, p 96

Lafargue E, Espitalie J, Marquis F, Pillot D (1998) Rock-Eval 6 applications in hydrocarbon exploration, production, and soil contamination studies. Inst Fr Pet 53:421–437

Masran TC, Pocock SA (1981) The classification of plant derived particulate organic matter in sedimentary rocks in J. brooks, organic maturation studies and fossil fuel exploration, London. In: 5th international palynological conference, vol 441. Academic

Mehrotra NC (2002) Palynology in hydrocarbon exploration: the Indian scenario. Mem Geol Soc India no. 61

Nwajide CS (2013) Geology of Nigeria's sedimentary basins. CSS Bookshops Limited, Lagos, p 565

Nwachukwu JI, Chukwura PI (1986) Organic matter of Agbada Formation, Niger Delta. Nigeria AAPG Bull 70:48–55

Oboh-Ikuenobe FE, Yepes O, Leg ODP (1997) Palynofacies analysis of sediments from the Cote d'Ivoire-Ghana transform margin: preliminary correlation with some regional events in the equatorial Atlantic. Palaeogeogr Palaeoclimatol Palaeoecol 129(3–4):291–314

Ojo OJ, Kolawole AU, Alalade B (2003) Paleoenvironment and petroleum source rock potential of the Enugu and Mamu formations, Anambra Basin Nigeria. Assoc Pet Explor

Ojo OJ, Ajibola UK, Akande SO (2009) Depositional environments, organic richness, and petroleum generating potential of the Campanian to Maastrichtian Enugu formation, Anambra basin, Nigeria. Pac J Sci Technol 10(1):614–628

Ojoh K (1990) Cretaceous geodynamic evolution of the southern part of the Benue Trough (Nigeria) in the equatorial domain of the south Atlantic: stratigraphy, basin analysis and paleogeography. Bull Centres Rech Explor Prod Elf Aquitaine 14:419–442

Okeke KK (2024) Depositional facies and palynofacies provenance reconstruction of the Danian Nsukka Formation, Southeastern Nigeria. Arab J Geosci 17(5):1–20

Okeke KK, Umeji OP (2016) Palynostratigraphy, palynofacies and palaeoenvironment of depositionof selandian to aquitanian sediments, southeastern Nigeria. J Afr Earth Sci 120:102–124

Okeke KK, Umeji OP (2018a) Palynofacies, organic thermal maturation and source rock evaluation of Nanka and Ogwashi formations in updip Niger Delta Basin, Southeastern Nigeria. J Geol Soc India 92:215–226

Okeke KK, Umeji OP (2018b) Oil shale prospects of Imo Formation Niger Delta Basin, southeastern Nigeria: palynofacies, organic thermal maturation and source rock perspective. J Geol Soc India 92(4):498–506

References

Okeke KK, Osterloff P, Ukeri P (2021) Sequence stratigraphical interpretation of the Paleocene to Miocene (Selandian–Aquitanian) palynofacies framework of the Niger delta basin, southeastern Nigeria. J Afr Earth Sc 178:104158

Okeke KK, Mode A, Anigbogu EC, Umeadi IM, Odu NJ, Maduewesi CO, Ulasi NA (2023a) Source rock potential, palynofacies depositional environment synthesis and structural traps of the Enugu Formation Southeastern Region, Nigeria. Arab J Geosci 16(5):1–17

Okeke KK, Umeji OP, Dim CP, Ekwenye OC, Ulasi NA, Uwakwe OC, Maduewesi CO (2023b) Depositional facies and palynofacies provenance of clastic deposits: insight from Paleocene strata in Southeast Region, Nigeria. Ir J Sci 47(1):73–90

Okeke KK (2017) Palynostratigraphy and granulometric assessment of paleocene to early miocene sediments. In: Awka–Onitsha Area, Niger Delta Basin, Southeastern Nigeria. MSc. thesis, University of Nigeria Nsukka, pp 1–240

Okeke KK (2023) Palynostratigraphy, palynofacies and paleoenvironment reconstruction of the Late Campanian to Danian strata of the Anambra Basin, Southeastern Nigeria [Ph.D Thesis]. University of Nigeria Nsukka, pp 1–344

Pearson DL (1984) Pollen/spore color "Standard," version 2, Phillips Petroleum Co., privately distributed. In: Travers A (ed) Paleopalynology, 2nd edn. Springer, Dordrecht, pp 581–613

Peters SW (1983) Littoral and anoxic facies in Benue Trough: Bull des Centre de Researches Exploration-Production. Elf Aquitane 7:361–365

Rayment RA (1965) Aspects of the geology of Nigeria. Ibadan University Press, pp 2–115

Reijers TJA (1996) Selected chapters geology, sedimentary geology, and sequence stratigraphy in Nigeria and three case studies a field guide. Shell Petroleum DeveloCompany of Nigeria Corporate Reprographic Services, Warri, p 197

Reineck HE, Singh IB (1980) Depositional sedimentary environments: with reference to Terrigenous Clastics. Springer, Berlin, pp 162–182

Schrank E, Mahmoud MS (2000) New taxa of angiosperm pollen, miospores and associated palynomorphs from the early Late Cretaceous of Egypt (Maghrabi Formation, Kharga Oasis). Rev Palaeobot Palynol 112(1–3):167–188

Short KC, Stauble AJ (1967) Outline of geology of Niger delta. Am Asso Petrol Geol Bull 51:761–779

Staplin FL (1977) Interpretation of thermal history from colour particulate organic matter: a review. Palynology 1:9–18

Stemmerik L, Christiansen FG, Piasecki S (1990) Carboniferous lacustrine shale in East Greenland-additional source rock in northern North Atlantic? Chapter 17, pp 227–286

Sykes R, Snowdon LR (2002) Guidelines for assessing the petroleum potential of coaly source rocks using rock-eval pyrolysis. Org Geochem 33(12):1441–1455. https://doi.org/10.1016/s0146-6380(02) 00183-3

Thomson BM (1982) Land plant source rocks for oil and their significance in Australian basins. Aust Explor Assoc J 22:164–170

Travers A (2007) Paleopalynology, 2nd edn. Springer, Dordrecht, pp 616–665

Tripathi SK, Divya Srivastava DS (2012) Palynology and palynofacies of the early Palaeogene lignite bearing succession of Vastan, Cambay Basin, Western India

Tyson RV (1993) Palynofacies analysis. In: Jenkins DG (ed) Applied micropalaeontology. Kluwer Academic Publishers, The Netherlands, pp 153–191

Tyson RV (1995) Sedimentary organic matter. In: Organic facies and palynofacies. Hapman and Hall, London, p 615

Umeji OP (2005) Palynological study of the Okaba coal mine section in the Anambra Basin, Southern Nigeria. J Min Geol 41(2):194

Umeji OP, Edet JJ (2008) Palynostratigraphy and paleoenvironments of the type area of Nsukka Formation of Anambra Basin, Southeastern Nigeria. Niger Assoc Petrol Explor Bull 20:72–89

Umeji OP, Nwajide CS (2007) Age control and designation of the standard stratotype of Nsukka Formation of Anambra Basin, southeastern Nigeria. J Min Geol 43(2):147–166

Umeji AC (2000) Evolution of the Abakaliki and the Anambra sedimentary basins, southeastern Nigeria. In: A REPORT Submitted to The Shell Petroleum Development Company Ltd, p 155

Varma AK, Hazra B, Chinara I, Mendhe VA, Dayal AM (2015) Assessment of organic richness and hydrocarbon generation potential of Raniganj basin shales, West Bengal, India. Mar Pet Geol 59:480–490

Venkatachala BS (1981) Differentiation of amorphous organic matter types in sediments

Welte DH, Tissot B (1984) Petroleum formation and occurance. Springer, New York, NY, p 699

Whitaker F (1982) Palynofacies investigation in the jurassic interval of the Norske shell well 31/2-4. Shell International Petroleum Maatschappij B.V., pp 1–14

Whitaker MF (1984) Palynological result of good 6407/9-1. A/S Norske shell exploration and production, pp 1–25

Zarboski PMP (1982) Campanian and Maastrichtian Sphenodiscid ammonite from Southeastern Nigeria. Bull Brit Mus Nat Hist 36:303–332

Zarboski PMP (1983) Campanian and Maastrichtian ammonite correlation and paleogeography in Nigeria. J Afr Earth Sci 1:51–63

Chapter 5
Tectonics and Structural Geology of Hydrocarbon Traps

Abstract Several natural gas flare and oil seepages uphold the source rock potential, structural trap synthesis and lateral facies change analysis of the studied Enugu Formation and the Danian units, for a detailed comprehensive structural trap style, hydrocarbon migration pathway from source rock to the reserves rocks for effective potential efficient hydrocarbon exploration and exploitation drive within the Anambra Basin. The study's sedimentary structures trapping mechanism and fluid migration pathway model are highlights of source rock potential, hydrocarbon and fluid production architecture of the earth's crust authenticated by the pore spaces and open and close joint system. The key structural traps of hanging-wall closures, footwall closures, half graben, horst, collapsed-crest-graben of conjugate fault array along with infilled and open joint structures within faulted and unfaulted sandstone sequences exhibits the structural hydrocarbon entrapment mechanisms of the Enugu Formations and Danian outcropping sediments. Primary and secondary hydrocarbon migration pathway in sedimentary rock pores and fracture sequences are the primary means of creating petroleum accumulations and reserves in the Anambra Basin. The study's oil seepages and natural gas flares reveal the formations' structural style elements as a petroleum system design for hydrocarbon entrapment and preservation model within the Anambra Basin. The structural style complexity, lateral and vertical facies changes, and hydrocarbon prospectivity authenticate the oil seepages and gas flare prediction, subsequent petroleum crude oil and gas drilling pathway and exploration and exploitation campaign in the Anambra Basin.

Keywords Structural traps · Oil seepage · Petroleum migration · Source rock · Primary migration and Secondary migration

5.1 Tectonics and Structural Traps Synopsis

Tectonics and structural geology clarification of sedimentary rocks uphold the systemic and dynamics of subsidence, folding and uplift substantiated by the evolution and dynamic creation history and timing of sedimentary basin origin in relation to the migration of hydrocarbons. The structural displacement of rocks along fault horizon is both vertical as inherent in normal faults and horizontal, concerning strike slip fault dynamics. The reverse faults exhibit fault system dynamics observed where the hanging wall is displaced or moved in a vertical direction/upwards relative to the footwall underneath the fault plane horizon. The key structural traps and series of joint architecture of the Enugu Formation and Danian strata show that the Anambra Basin depositional area is a typical domain with high horizontal stresses of the earth as a result of the diverging plate movements.

Traps are classified into two with respect to the type of structure that produces them namely: structural traps and stratigraphic traps. Structural traps are products of structural deformation substantiated by rocks' folding, doming or faulting. Stratigraphic traps are triggered by primary features of sedimentary rock sequences formed in the absence of structural deformation events of faulting or folding. Pinch-out stratigraphic structures, definitive of sandstones pinching out in shales depicts the resultant effect of primary changes in sedimentary rock facies. A cumulative sedimentary structural style substantiated by stratigraphic and structural traps depicts a combination trap or strati-structural trap prototype (Table 5.1).

The fault plane mechanics of fault trap is amongst the structural style that traps the oil and hinders its further upward migration. This illustrates fault planes of fault trap as sealing traps for vertical flow or upward migration of petroleum oil and gas that mimics the function of a barrier along with cap rock for reservoir rocks.

However, the juxtaposition of any reservoir rock against sandstone or permeable rocks makes the fault an impermeable structure that mitigates flow across the fault plane. The mudrock units at the Enugu and Ikpankwu section shows that the reservoir rock is consistently faulted against a tight shale and other mudrock facies which subjected the fault to a sealing type fault. In light of this, fractures and a series of key fault prototypes on the shales and other mudrock facies of the studied formations reveal an infilling of brown and whitish consolidated sediment material similar to quartz cementation, capable of rendering the fault plane a less permeable structure. This may be due to temperature related diagenetic changes within > 3 to 4 km depth during the deposition and deep burial of the Anambra Basin sediments (Figs. 4.5, 4.6 and Table 5.1). These diagenetic changes of the fault plane have been reported in the structural and petroleum geology analysis in Permian Zechstein salt in Germany and Denmark, Kimmeridge Clay (Upper Jurassic) Southern Ferriby, Yorkshire, England, the North Sea basin, and mid Norway (Haltenbanken) along with the Barents Sea and other areas of the world (Bjorlykke 2010).

A structural fault plane is a sealing morpho-strata. Although this event is not always articulated in all fault planes as shown in Fig. 5.1 this structure is capable of stopping oil and gas consistent and frequent upward or downward migration. Traps comprise a porous reservoir sandstone rocks overlain by mudrock facies tight (low permeability) rocks that negates the migration of oil or gas petroleum through the pore spaces of rock.

5.1 Tectonics and Structural Traps Synopsis

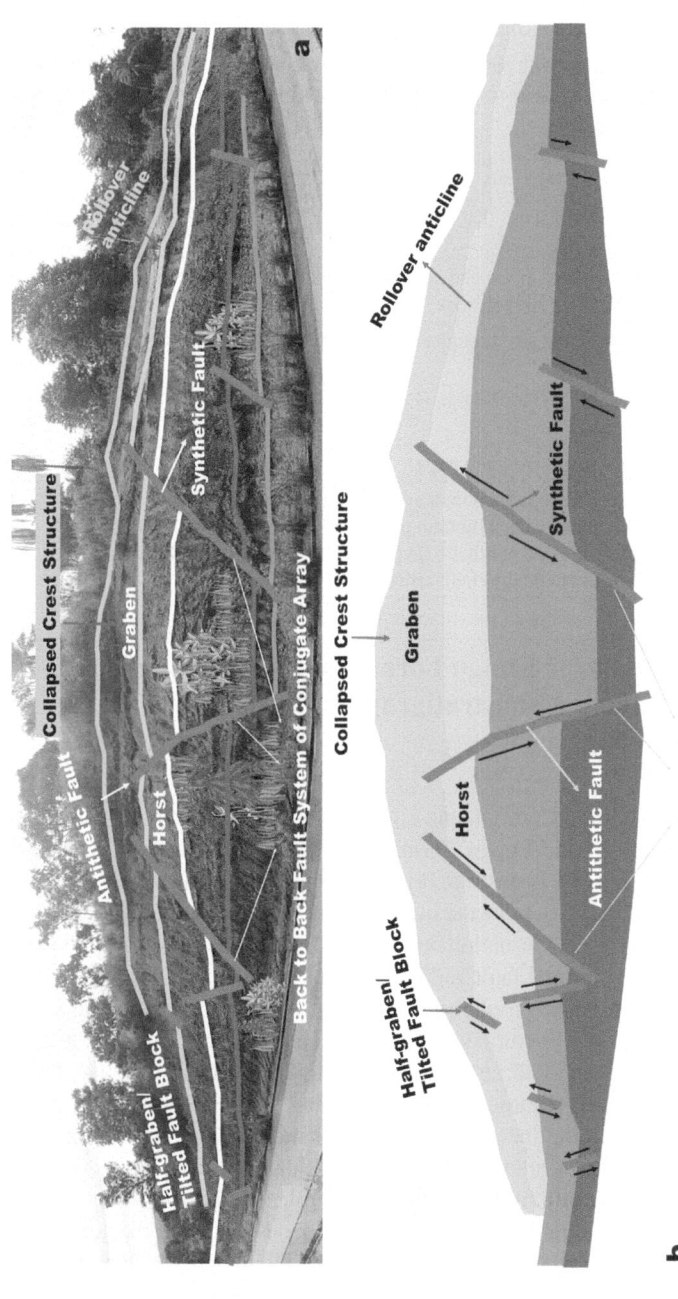

Fig. 5.1 **a** Outcrop section of the Enugu Formation at Neke Isi-Uzo road cut off-Opi Abakpa-Nike Bypass, along Ugwuogo Nike–Ikem road portraying conjugate fault system with pronounced normal fault arrays on the thick shale with siltstone interbeds revealing the hanging wall and footwall blocks of the synthetic and antithetic fault plane layers in each conjugate faulted bed with other structural attributes of the regional fracture system on outcrop basis. **b** Fracture system graphic model of the conjugate fault system of the Enugu Formation, at Neke Isi-Uzo road cut off-Opi Abakpa-Nike Bypass, along Ugwuogo Nike-Ikem road Enugu State, southeastern Nigeria showing series of synthetic and antithetic faults, horst and graben and other structural dimensions of the faulted outcrop

Table 5.1 Geometric classification of fault arrays and petroleum migration terminology

Terminology	Definition
Joint array	Any group of systematic or nonsystematic series of closed or open joints
Systematic joints	A group of joints with parallel or subparallel relationships to one another, with a relatively evenly average spacing quality similar to one another in a physical geology region
Nonsystematic joints	A joint array with an irregular spatial spreading with no parallel joints dynamism and a relatively nonplanar design
Joint system	A series or sets of joints that are geometrically related in a formation, basin or region
Migration	The flow of petroleum/oil and gas from mudrock facies source rocks through carrier bed to the reservoir rock
Primary migration	The flow of oil and gas through and out of the mudrock facies source rocks
Secondary migration	The flow of oil and gas through wider pores spaces to the trap
Dismigration	Loss of oil and gas from the traps
Anticlinal structures	Encompasses the anticlinal fault system characterized by several faulted blocks of synthetic (Dipping fault plane within the direction of the regional stratigraphic dip) and antithetic (Dipping faults plane against the direction of regional stratigraphic dip) fault arrays

5.2 Fault Systems of the Enugu Formation, of the Anambra Basin and Danian Lithostratigraphic Units, Southeastern Nigeria

The palynofacies depositional environment, source rock potential and sedimentary rock structures hallmark of the Enugu Formation within Neke Isi-Uzo, Amagu, Aguabo outcrop by the Aguabo pedestrian bridge road cut, and Ozalla/Four-Corner road cut outcropping sections designate the studied faults system. The Danian horizon at the Ikpankwu and the Enugu Formation outcrops expatiated the complexity of the several structural styles of normal fault and growth fault system, stratigraphic traps along with quality trap mechanism of the formations for hydrocarbon accumulation (Figs. 3.1 and 5.1). The structural stratigraphic nomenclature of the Anambra Basin group of tectonically related faults is called a fault system in this research. However, faults array is also regarded as a synonymous jargon of van de Pluijm and Marshak (2004) concerning the geometric classification of fault. The sedimentary rock significant structures renowned within the lithostratigraphic units of the Enugu Formation in the vicinity of Neke Isi-Uzo along Ugwuogo-Nike-Ikem road and the Danian outcropping unit at the Ikpankwu section, designated as the key fault system prototypes of the formations are typified by footwall, hanging-wall and anticlinal structures with sequence of closed and open joint structures (Fig. 5.1). Wedge/pinch-out structures and strati-structural/combination trap were also recorded in the

5.2 Fault Systems of the Enugu Formation, of the Anambra Basin … 71

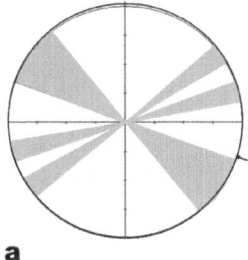 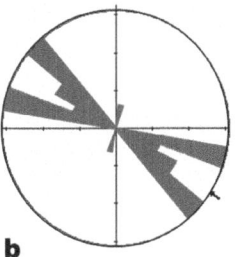

a

Axial (non-polar) data
No. of Data = 5
Sector angle = 10°
Scale: tick interval = 5% [0.3 data]
Maximum = 20% [1 data]
Mean Resultant dir'n = 107-287

b

Graphical representation of joints
from the Enugu Shale at the
Aguabo pedestrian bridge outcrop
of the Enugu Formation

Fig. 5.2 **a** A Rose diagram plot of the Enugu Formation joints at the Aguabo pedestrian bridge outcrop section within the Enugu vicinity, SE Nigeria. **b** A Rose diagram (frequency-azimuth) plot of Enugu Formation joint fractures at Ozalla outcrop unit

course of this research. These fault closure collections are discussed as conjugate arrays, normal fault systems, joints and series of stratigraphic traps. The normal fault system and joint structure arrays in the studied formations of the Anambra Basin (Fig. 5.1) exhibit a regional faults standard model formed in rifts that is regarded as belts with effective lithospheric extensional force in structural geology. These structural archetype styles are (i) anticlinal structures, (ii) growth faults, (iii) footwall and (iv) hanging wall.

5.2.1 Anticlinal Structures

The structural attributes of an anticlinal fault system are characterized by several faulted blocks of synthetic (A dipping fault plane within the direction of the regional stratigraphic dip) and antithetic (A dipping faults plane against the direction of regional stratigraphic dip) fault arrays. Anticlinal traps are capable of forming in association with faulting. In the absence of growth fault mechanism in the Neke Isi-Uzo fault plane, the anticlinal traps form in association with faulting substantiated by roll-overs anticline of growth faults and thrust zones terrane. Synthetic faults are fault systems with a similar dip direction as the major fault in the regional stratigraphic direction. In contrast, antithetic fault array are faults dipping against the direction of the major fault within the regional stratigraphic dip direction array (Fig. 5.1a and b). The dimensions of the structural fault is dipping 51° with a dip direction of 320° within the northwest direction (Fig. 5.2). The numerical value of

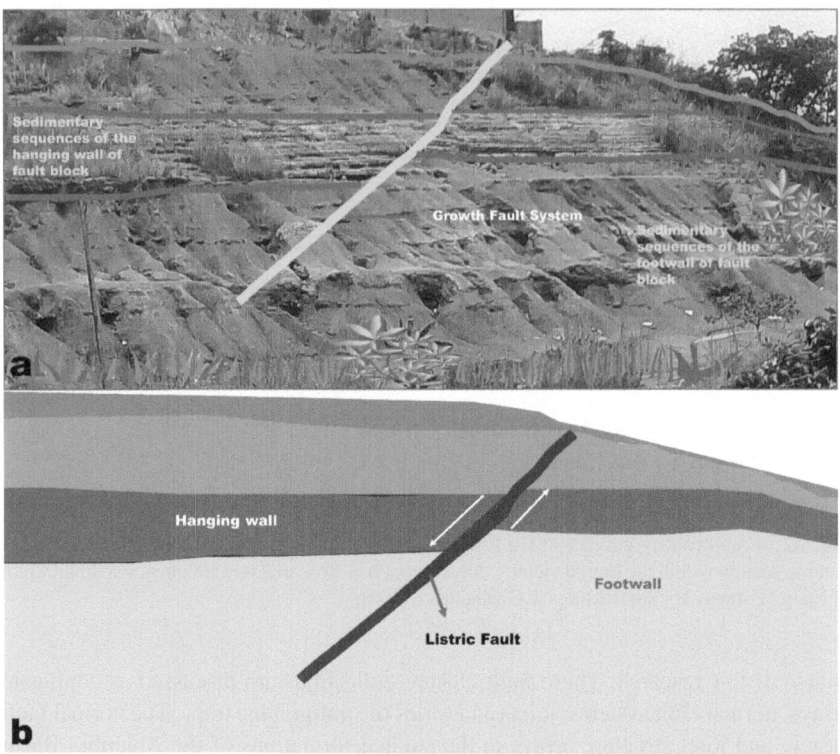

Fig. 5.3 a The Aguabo outcrop section portraying a heterolithic interval with systematic growth fault in the Enugu Formation with rollover structures on both hanging wall and the footwall block exposed near Onitsha road Aguabo pedestrian bridge in Enugu, SE Nigeria. b A graphic structural up-thrown and down-thrown block components of the faulted section in the Enugu Formation at the Aguabo pedestrian bridge by Enugu-Onitsha roadcut, Enugu area, SE Nigeria

the regional dipping style of the Enugu Formation of the Anambra Basin was previously reported by Amogu et al. (2010), Anigbogu (2013), Okeke et al. (2023). In this research, horsts, half graben and grabens are common products of synthetic and antithetic fault system interaction of the anticlinal structures generated in rift systems (Fig. 5.1). However, the studied outcrop sections at Neke Isi-Uzo along Ugwuogo-Nike-Ikem road shows that the combination of conjugate fault array gave rise to down-warping inherent in the dipping of the two faults towards each other within the crestal region, forming a collapsed-crest or graben. This structural attribute is synonymous to graben system. The anticlinal fault array and normal fault system mechanics is the reported systematic functional characteristic features of the seismic and outcropping structural traps of the Niger Delta Basin and Anambra Basin where the fault blocks are juxtaposed against the downthrown and upthrown wall sections (Short and Stauble 1967; Germeraad et al. 1968; Adesida et al. 1977; Doust and

5.2 Fault Systems of the Enugu Formation, of the Anambra Basin … 73

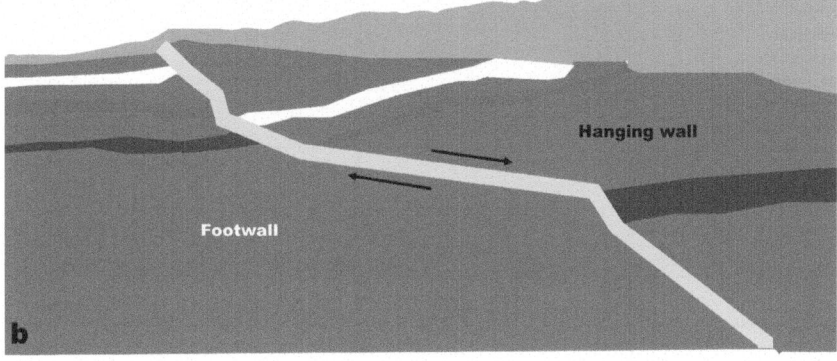

Fig. 5.4 **a** An Outcroping unit of the Enugu Formation, Anambra Basin at Amagu by Enugu-PortHarcourt roadcut showing normal fault system on the pronounced thick shale strata with siltstone interbeds. Geologists, positioned on the hanging wall closure and footwall block of the regional fracture. **b** Fracture array graphic model of the regional normal fault showing the down-thrown (hanging wall closure) and up-thrown (footwall closure) block visible at Amagu vicinity within the Enugu Formation along Enugu-Port Harcourt roadcut, southeastern Nigeria

Omatsola 1990; Stacher 1995; Amogu et al. 2011; Dim et al. 2020; Okeke et al. 2023).

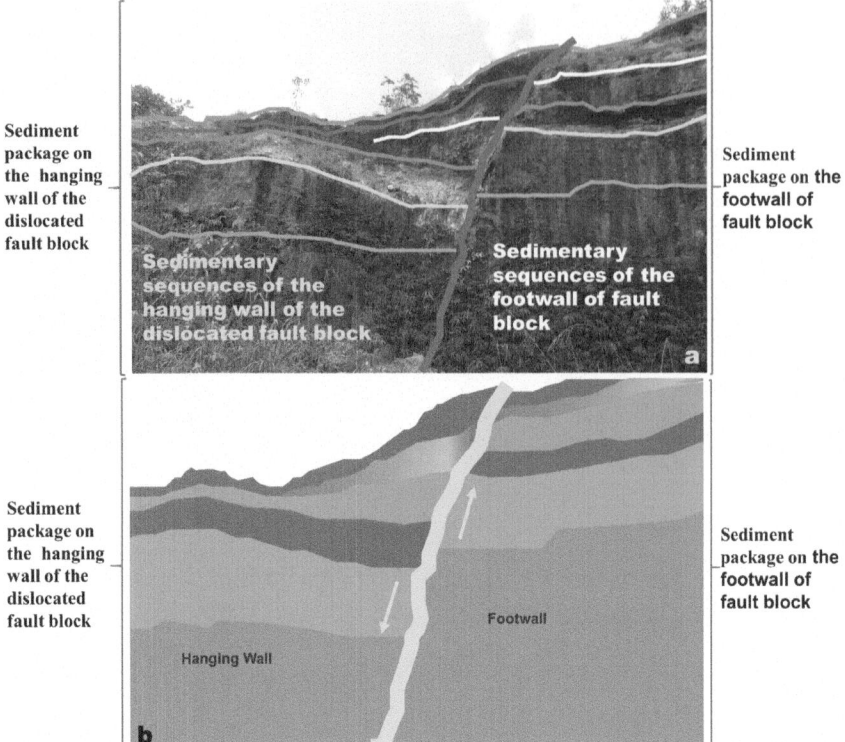

Fig. 5.5 a Normal fault system of the Danian section of southeastern Nigeria showing the hanging wall closure and footwall closure elements at Ikpankwo quarry along with the deleaneated sediment packages on the (dislocated) down-thrown block (hanging wall) and up-thrown block (footwall) of the fault block. **b** Pictographic modeled normal fault illustration of the trapping components of the Danian unit at the Ikpankwu outcropping section off Enugu–Port Harcourt road, southeastern Nigeria showing the sediment packages on the hanging wall closure (dislocated) and footwall closure layers of the fault block fracture

Generally, the evidence of the uplifted sediment packages at Neke Isi-Uzo is supported by two faults dips, away from each other producing a back-to-back structure and forming horst block structures. This substantiates normal fault trending features and relative similarity along with conjugate fault array (Fig. 5.1). The anticlinal structures with unique synthetic and antithetic faults system exhibit good entrapment for fluids and liquid hydrocarbon accumulation potential by means of simple or faulted closures. The hydrocarbon bearing reservoirs in the Niger Delta and Anambra basins are often found to be trapped within the anticlinal structures (Doust and Omatsola 1990; Dim et al. 2020; Okeke et al. 2023).

The structural style mechanism of the Enugu Formation at Neke Isi-Uzo shows that oil traps are also formed in anticlines within the upper side of the fault plane. The anticline measures/pattern in the oil traps are products of a series of salt domes

5.2 Fault Systems of the Enugu Formation, of the Anambra Basin …　　　　　75

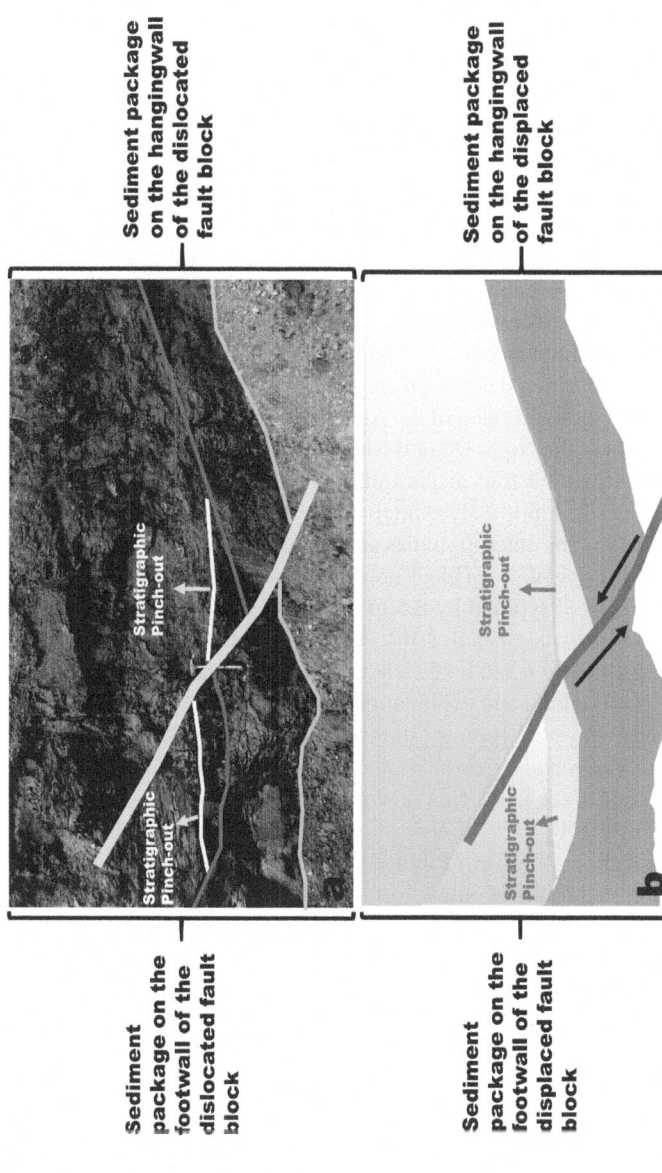

Fig. 5.6 a Strati-structural trap of normal faulting model of thin siltstone lens component of the outcropping sandstone unit of the Enugu Formation, Anambra Basin at Ozalla/Four-Corner Junction along Enugu-Port Harcourt road cut exposure, southeastern Nigeria showing the faulted siltstone block and shale block. (Geologic hammer scale = 30 cm). **b** An interpreted schematic combination trap model normal fault showing faulted siltstone/sandstone lens exposed at Four Corner, Ozalla outcropping road cut section along Enugu Port-Harcourt road, Enugu southeastern Nigeria

and anticlinal folds or basement high (Knut 2015). These are formed by diapirism or other geologic processes capable of four-way closure or formation of stratigraphic rock closures in all directions (Fig. 5.1).

5.2.2 Growth Faults

Growth faults are products of gravity-sliding within curved (listric) fault plane horizons that are usually typified by sedimentary sequences deposited in rapid sedimentation due to the hydrodynamic and saltation of sandstone within the relatively high energy marine deltas depositional space. The growth fault system at the Aguabo pedestrian bridge road cut section along Enugu-Onitsha road shows that the faulting mechanism of the formation is relatively active during the rapid sandstone sedimentation which revealed the thickest downthrown section of the sedimentary rock horizon deposited due to the deltaic influences of the Enugu Formation within the outer neritic water realm (Fig. 5.3). This is a defined prototype of the rollover anticlines in the Cretaceous sedimentary rock secession of the Anambra Basin similar to the series of rollover anticline design of the Niger Delta Basin, southeastern Nigeria. The growth fault mechanism and structural trap display of these basins were formerly documented (Dim et al. 2020; Nwajide 2013; Umeji 2000; Anigbogu 2013; Okeke et al. 2023; Amogu et al. 2011). The outcrop studies and numerical values of the measured sediment thickness on the downthrown block are denoted as 10.7 m whereas the thickness on the upthrow block is typified by 8.5 m with a previously validated growth index range value of 1.35 (Amogu et al. 2011; Anigbogu 2013).

The structural trap dimension and nomenclature of "growth fault" were coined in the early days of oil exploration and exploitation era in the absence of seismic data acquisition technology. Physical geology studies of the era revealed that geologic units of the faulted horizon had "grown" in thickness in wells and outcrops in the downthrown section side of the fault. This shows that the displacement of the faulted beds decreases upwards and increases downwards along the fault plane. However, growth faults are usually of low permeability that substantially bring on reduced pore water circulation within the geologic formations of sedimentary basins where the under compacted clay, that can develop into clay diapirs that occurs in relation with growth faults (Bjorlykke 2010) can be ascertained.

5.2.3 Footwalls

Footwalls of regional normal fault systems are products of rift belts with lithospheric extensional force along passive margins. This is because the Andersonian structural fault concept buttressed the geometry of a fault system as products of regional stress conditions of the earth's horizon that caused faulting stress and footwall mechanics. The footwall design of the faulted blocks of the study is qualitative,

5.2 Fault Systems of the Enugu Formation, of the Anambra Basin ...

describing an upthrown fault block of normal fault within a conjugate fault system. These are an upthrown fault or un-displaced normal fault block in a structurally deformed lithospheric setting inherent in a listric fault structural array. A typical hanging wall design is shown in Figs. 5.1, 5.3, and 5.4, recorded in all the studied outcrops. This is consistent with the sedimentary structural dynamics of growth faults where the sediment packages thicken on the downthrown block. The juxtaposition of these fault blocks against the walls of the downthrown section exhibits the characteristics of a sealing fault block. It provides a systematic entrapment structural style for fluid (Figs. 5.1, 5.3, and 5.4) and liquid petroleum accumulation, which is productive in petroleum geology, particularly when reservoir rocks flank a non-reservoir rock package. This is the characteristic highlight of the seismically exposed structural traps style of the Niger Delta Basin and Anambra Basin (Short and Stauble 1967; Germeraad et al. 1968; Adesida et al. 1977; Doust and Omatsola 1990; Stacher 1995).

5.2.4 Hanging Wall

The hanging wall design of the faulted blocks of the Enugu Formation along the Neke Isi-Uzo, Amagu, Aguayo and Ozalla Four Corner outcrops and the Danian section of the Nsukka Formation at Ikpankwu exemplify a normal fault system, which is relatively apparent in conjugate fault array architecture, where the downthrown fault block of structurally distorted setting is also associated with listric (planar) faulting or growth fault. This is also regarded as a distortion with distinct fault bends with a visually inclined relationship of diverse sets of faults at an angle of $\pm 60°$ (Knut 2015). These fault blocks are considered as rollover anticlines or tilting of overlying fault blocks and are juxtaposed in opposition to the footwalls. In the Neke Isi-Uzo fault system, the primary top surface of the hanging-wall block that tilts towards the fault plane to make a depression is called a half-graben whereas the two adjacent normal faults-block system that dips towards one another when the fault-bounded block between them drops down is the graben. The conjugate fault array examined at the Neke Isi-Uzo vicinity road cut section and other non-conjugate fault system are all typical series of normal fault system which exhibits conditions where the regional stress $\sigma 3$ (sigma) is horizontal and trends at a high angle to the trend of the system (Figs. 5.1, 5.3, 5.4, and 5.5). The geometry dynamics of rollover anticline structures along with liquid hydrocarbon distribution and other fluids reservoir quality of the structures were also illustrated in the seismically inclined subsurface Niger Delta Basin (Nwajide and Reijers 1996). This was also illustrated in the structural fault system of the outcropping units of the Anambra Basin (Okeke et al. 2023; Amogu et al. 2011; Anigbogu 2013). The hanging wall of a normal fault has the potential for sealing and entrapment of fluid, which suggests that it is impermeable to fluid migration across the fault, indicative of a sealing fault trap. However, in reservoir scale petroleum geology hydrocarbon generation analysis of sedimentary basin, a typical sandstone reservoir unit are flanked by non-reservoir source rock shale demonstrated in this research (Figs. 5.1, 5.2, 5.3, 5.4, and 5.5).

5.2.5 Stratigraphic Traps

Stratigraphic traps are geologically initiated, produced and authenticated as imprints of facies variation or unconformities layers without tectonic deformation. Stratigraphic traps inherent in the study are obvious within the lithostratigraphic units of the Enugu Formation in the vicinity of Ozalla/Four Corner road cut (Fig. 5.6). Other series of stratigraphic traps structural styles within the formations of the Anambra Basin were reported by Dim et al. (2020) and Anigbogu (2013). The strati-structural trap configuration model reported in this work is systematic non tectonic and tectonic traps inherent in combination trap style for fluid accumulation. These strati-structural/combination trap configurations exist as (i) wedge/pinch-out structures and (ii) strati-structural/combination traps.

5.2.5.1 Wedge/Pinch-Out Structures

Enugu Formation wedge-like and pinch-out stratigraphic structures exposed at the lower unit of the formation were encountered at Ozalla/Four Corner road cut outcropping section (Fig. 5.6). Relatively porous and permeable very fine-grained to siltstone sandstone units which pinch out up-dip and downdip in less permeable rocks, shale replicate the pinch out stratigraphic style of the study (Fig. 5.6). The attributes of the stratigraphic trap are products acquired on the flanks of fluvial channel fills owing to thinning out of sediment packages during deposition. This stratigraphic trap style is an authenticated entrapment mechanism for fluid accumulation (Dim et al. 2020; Knut 2015).

5.2.5.2 Strati-Structural/Combination Trap

A combination trap of Enugu Formation at Ozalla/Four Corner road cut outcrop emerged as a result of both tectonic deformation and facies variation or unconformities as pinpointed in Fig. 5.6. This faulted sandstone lens was detected at the formation's basal unit, creating two juxtaposed wedge-like/pinch-out structures typical of both stratigraphic and structural traps (strati-structural/combination trap). This is a typified tectono-sedimentary process of depositional sequences. Oil seepages and series of gas flare from borehole drilling mechanics (Caritas University, Ugwogo Nike and Amorji) are products of pinchouts observed at Ozalla which explained the effectual shallower oil and gas traps mechanics in the Enugu area of the Anambra Basin. These pinchouts depict a series of revitalization of tectonic movements on the fault systems displaced and juxtaposed fluctuating marine and non-marine continental deposits through the strata. This projects the systematic favorable formation of jointly stratigraphic and structural traps inherent in the studied Enugu Formation and Danian sections. The structural design of combination trap triggers secondary fluid migration and hydrocarbon accumulation inherent in quality trap mechanism.

5.3 Fault Systems and Petroleum Migration Pathway of Enugu Formation of the Anambra Basin and Danian Units in Southeastern Nigeria

Natural gas flare and oil seepages, basin modelling along with conventional and unconventional petroleum exploration and exploitation prospects of the Anambra inland basin triggered the synthesized sedimentary depositional process disclosure, structural style of the fracture systems and petroleum geology system of the studied Enugu Formation and Danian stratigraphic units in southeastern Nigeria. Petroleum migration physical mechanics between the source rock (Primary migration) and first entrapment (secondary migration) is a factual instrumental measure for basin modelling and petroleum exploration and exploitation campaign (Dim et al. 2020; Schowalter 1979; England et al. 1987; Clarke et al. 1988; Ungerer 1990; Hindel 1997; Okeke et al. 2023). The oil migration mechanism in this work harnessed synthesized geological techniques geared towards a unified structural migration pathway prospects assessment, organic matter geochemical analysis for a better organic thermal maturation results and hydrocarbon expulsion stage along with other physical petroleum principles itemized herein. Several petroleum migration physical mechanics model via source rock (Primary migration) to reservoir rock (secondary migration) were articulated in basin modelling and petroleum exploration (Ungerer 1990; Hindel 1997; Clarke et al. 1988; Dim et al. 2020; Schowalter 1979; England et al. 1987; Hunt 1979; Tissot and Welte 1984; Okeke et al. 2023) to understand the quantity and quality scale dimensions of micropores, macropores and fractures in source rocks. Primary migration, is the flow of hydrocarbon out of the source rock while secondary migration depicts flow of petroleum from the source rock to the reservoir rock via structural style or seep up to the surface. The major direction of primary migration of fluids generated in the source rock can flow in a vertical or horizontal pattern (Magara 1980).

Diffusion is amongst the principal mechanism of primary migration where oil and gas are diffused from source rock micropores and mesopores to macropores and fractures depicting the source rock to reservoir rock oil conduit flow design. Diffusion is indicated as the primary migration mechanism at play, after petroleum generation before the process of a pressure-driven fluid flow system (Mann 1994). This illustrates that oil and gas are capable of leaking (migrate) from the sandstone reservoir to an advanced trap and to the surface. Primary hydrocarbon migrations stand out as principal mechanics in the petroleum accumulation build-up. Hydrocarbon gas adsorption–desorption and style of fluid flow are dynamic resultant effect of quality pore structures of carbonaceous shales (Loucks et al. 2009).

The source rock sedimentary succession of the formations consists of dark-grey, black and carbonaceous, sideritic, fissile, micaceous and non-fossiliferous shale lithofacies rock unit (Figs. 3.1, 3.2, 3.3, 3.4, and 3.5). The sedimentary rock attributes designation of the lithofacies units, facies variation or unconformities, rock texture and bedding planes of sedimentary rock units primarily regulate the possible fluid migration pathways and water fluids through the primary interparticle pore spaces of rocks. The low permeability rate inherent in nanodarcy level illustrated small particle fabrics as the typical highlight of carbonaceous shales (Singh et al. 2021). The dominant shale, mudstone and siltstone source rock facies of the study are the potential pathways for the migration of the hydrocarbon fluids which are products of a major mixture of Kerogen Types II and III which is oil and gas prone (Fig. 4.8). Hydrocarbon oil and gas accumulations are of organic origin and developed as systematic products of organic matter in sediments. The low hydrogen index (HI) and vitrinite reflectance values of this study portrays the prevalence of terrestrially derived Type III kerogen with the potential to generate gas (Ojo and Akande 2009). This substantiates the oil seepages and gas flares generated from the petroleum reservoir of the study, as a result of the structural hydrocarbon entrapment mechanism (Fig. 5.7). Buoyancy and groundwater flow are the two forces responsible for petroleum flow from source rock to traps (Hindel 1997).

However, a series of microfractures and macrofractures inherent in faults and joints are characteristic features of first entrapment (secondary migration) pronounced in joints, normal fault, growth fault along with the synthetic and antithetic structural architecture of the studied formations (Figs. 5.1, 5.2, 5.3, 5.4, 5.5, 5.6, and 5.7). The open and close joints in the Enugu Formation and Danian units

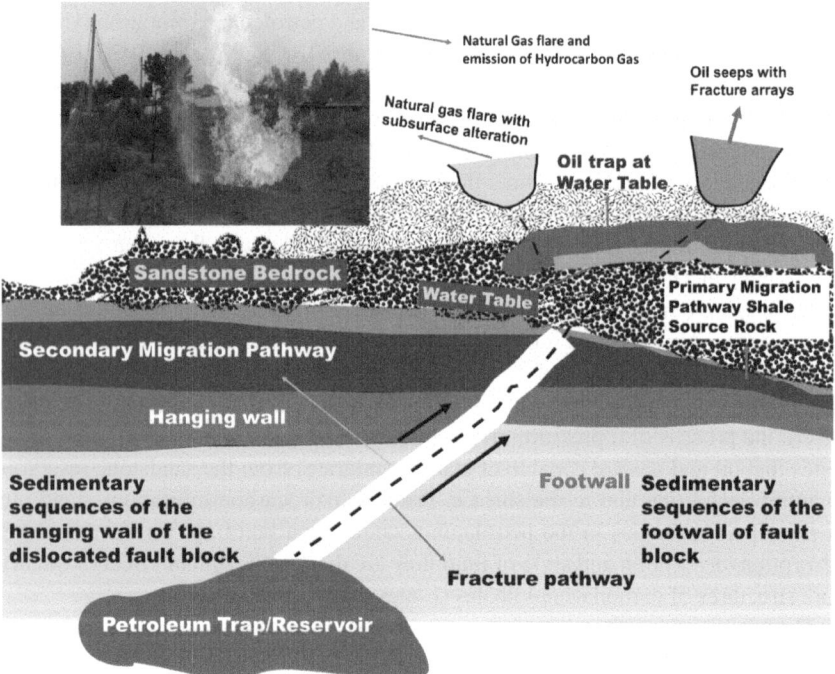

Fig. 5.7 Schematic modeled fracture system of the illustrated fluid/hydrocarbon seepage processes and dispersal mechanisms of primary and secondary migration pathways of the Enugu Formation petroleum system, Anambra Basin southeastern Nigeria

5.3 Fault Systems and Petroleum Migration Pathway of Enugu Formation ... 81

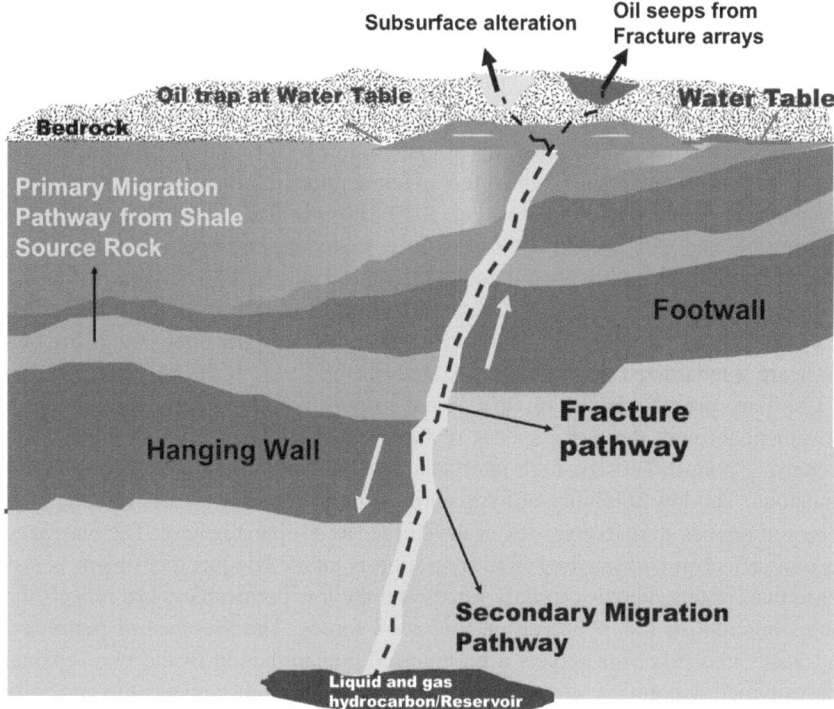

Fig. 5.8 Pictographic modeled normal fault system of the Danian horizon of the study revealing schematic design of fracture pathway, seepage processes and dispersal mechanisms at Ikpankwu quarry showing sediment package on the hanging wall and footwall of the fault block

are due to low permeability of the mudrock facies and the resultant increase of the fluid pressure in water and petroleum phases along with subsequent fracture of tight shales and systematic widening of the petroleum migration route. The hydrocarbon migration mechanics substantiates the upwards migration component and downward migration part of petroleum from a source rock into a sandstone or limestone reservoir rocks via oil conduits (Fig. 5.7).

The degree of oil migration is the product of the intensity of hydrocarbon generation in the shale source rocks. In contrast, the rate of temperature over time due to burial depth and geothermal gradients is the determinant factor necessary for source rock maturation system (Fig. 4.7 and 4.8). However, oil decomposes or is cracked into gas through deep burial over a long period of time in the earth's history and petroleum geology research system. The integrated depositional environment interpretation processes of depositional facies, rock textures and bedding plane characteristics of the sedimentary strata are regarded as the key parameters to consider in potential fluid migration pathways and other hydrocarbon fluids analysis across the primary interparticle pore spaces of the rocks. The consolidated sandstone, siltstone, and heterolithic rock units are possible reservoir rocks of the hydrocarbon system in the studied Enugu Formation and Danian sediments (Figs. 3.1, 3.2, 3.3, 3.4, and 3.5). Complex pore structures and extremely low permeabilities of organic-rich shales are the perceived challenges in hydrocarbon extraction and migration with increased temperature in petroleum pyrolysis. This pore pressure and low permeability indices of mesopore and micropore structures in shale source rock from low-pressure N_2 and CO_2 gas adsorption has shown the effect of temperature thermal treatment in migration pathway delineation (Chandra et al. 2020; Hazra et al. 2023). Results from low-pressure N_2 and CO_2 gas adsorption indicate that shale mesopore and micropore structures are significantly altered due to thermal treatment at higher temperatures.

In the migration of petroleum, capillary forces and permeability studies, two phases of fluid accumulation and migration pattern of the water phase and petroleum phase are standardized in reservoir rock analysis.

Capillary pressure is the resistive force responsible for little or no petroleum movement through the pores spaces of the rock. The forces field increases with decreasing pore-throat size, high interfacial tension, and pronounced wettability of sandstone. The low solubility of hydrocarbons in water authenticates its migration from source rock to reservoir rock or the surface as a separate phase. The buoyancy peculiar effect in hydrocarbon fluid migration is validated since petroleum is less dense than water whereas capillary forces and the low permeable rocks reflects the forces inherent in low hydrocarbon migration forces. The presence of petroleum and water in a reservoir depicts a pronounced contamination of the two separate phases which denoting a relatively economic loss. This substantiates the generally accepted knowledge that oil migrates as a separate phase in petroleum geoscience. Hydrocarbons exhibit solubility values of less than 1 ppm in the presence of water due to low pressure (Knut 2015). It is important to note that oil is lighter than water but the former migrates through reservoir rocks' pore spaces while high capillary resistance is noted for distinct oil dropslet in a water-wet rock.

The oil leakages and several natural gas flares in the Enugu Formation suggest that oil droplets must overcome the capillary forces to seep through pores (pore throat). However, the little pore spaces inherent in the fine-grained sediments/mudrock facies exhibit a barrier to additional oil migration due to the capillary forces. The incessant gas flare of the formation results from the small gas molecules which gradually diffuse via very small pores leading to slow gas seepage from the shale source rock that triggers and develops tight seals for oil. However, migration of oil from source rock (Primary migration) to reservoir rock (secondary migration) negates small discrete droplet movement (migration), but migrates as a continuous string of oil in the presence of diverse pores spaces stuffed with oil rather than water in high oil-saturated phase (Fig. 5.8).

The quality of cap rock slightly negates 100% effectiveness in preventing upward flow of hydrocarbons, but cap rocks allow water permeability and impermeability of oil and gas due to capillary force resistance in small pores of rocks. However, the mechanism of oil and gas migration in mudrock is similar to the oxygen diffusion in the clay-rich sediments where the pore space of clays is closed in the water system, resulting in the downward very slow diffusion of oxygen in fine-grained sediments of the seafloor. Coarse-grained sediments aerobic breakdown via oxygen diffusion is relatively more effective than in fine-grained strata.

The natural gas flare at Caritas University at 6° 30′ 27″ N and 7° 34′ 35″ E (Okeke et al. 2023) along with Ugwogo Nike and Amorji Nike at 06°30′29.1″ N and 7°34′23.7″ E (Nosike 2022) within the type area of the Enugu Formation and series of oil seepages reported in the Anambra basin initiated the wholistic attention of transport of hydrocarbons and fracture system from the carbonaceous shale source rock to the reservoir rocks. The natural gas flares from the Enugu Formation illustrates the surface dispersal petroleum efflux/seeps category of Clarke and Cleverly (1991). However, lack of migration in micropores, mesopores and effective diffusion pattern in oil and gas flow to macropores and structural trap arrays are the main primary migration pathway mechanics during diffusion of liquid and gas hydrocarbon in source rocks (Mann 1994; Okeke et al. 2023).

The sedimentary rock hydrodynamics and sedimentary environments control the distribution of reservoir rocks and primary composition of sediments, altered by thick overburden due to diagenesis sandstone depositional facies during burial. Diagenetic processes of rocks control diverse changes in porosity (compaction), permeability, and other physical and chemical attributes of velocity obtainable in both sandstone and limestone reservoirs rocks of sedimentary basins. However, reservoir rock's quality and excellent properties are substantiated by synthesized examination and interpretation of sedimentology (depositional environments) of reservoir rocks, sediment composition (provenance), diagenesis and the structural geology dimensions of geology rocks.

5.4 Kinematics of Faults and Joints in the Enugu Formation of the Inland Anambra Basin and Danian Units of Southeastern Nigeria

The structural style dynamics of the Enugu Formation at Aguabo pedestrian bridge, Neke-Isiuzo, Amagu and Ozalla Four Corner outcrops along with the Danian unit of the outcropping section at Ikpankwu exhibit a pronounced series of growth faults, normal faults and systematic and nonsystematic joint fracture system displayed in the pictograph in Figs. 5.1 and 5.3. These key fault porotypes of the Anambra Basin are authenticated by the displacement of the heterolith units at the mid-section of the Aguabo pedestrian bridge outcrop unit, and the displacement of siltstone and shale inter-beds within the mudrock unit for the Ozalla/-Four Corner, Neke-Isiuzo, and Amagu along with the Ikpankwu outcrop mudrock and sandstone facies fracture system structural styles. The kinematics of faults and joints of the study was assessed to understand the structural and sedimentary rock relationship, paleostress direction and stress field rotation in the formations and appraisal of fault system (Fig. 5.2). The synthesized assessment of these kinematic factors in oil and gas hydrocarbon migration pathways and fault system exposé buttressed the oil migration route and reservoir quality. The kinematic events visible in the structural style of the Anambra Basin will demonstrate the magnitude of the paleoestrees direction and stress field rotation as a regional or local compressional and extensional tectonic events. This is because the analysis, and interpretation of kinematics of sedimentary structures such as faults and joints on outcrop sections are exceptionally important in the appraisal of tectonic events, hydrologic application, geomorphologic studies and other geologic structures in basin modelling array (Figs. 3.5 and 1.2) .

The numerical orientation data from joint kinematic analysis projects vital result on the orientation of stress field events at the time of rock failure. The joint development processes (fracture, crack) causes drop/low pressure in fluids, because the generation of joints and faults creates spaces for fluid migration while the connectivity of joints and pores in rock is vital for the permeability network inherent in quality fluid migration. The structural style of faults and joints was examined using proper measurements of the orientation/trends using a geologic compass and a Global Positioning System (GPS). Joint and fault dataset were obtained from studied outcrop sections while the principal stress δ1, the intermediate stress δ2 and the minimum stress δ3 were generated with and plotted with the aid of Georient software (Fig. 5.2).

The data were sourced from the structural style of the studied outcrops of the area, revealing a systematic joint set within the NNW to NNE coordinate for the Enugu Formation and Danian units. Similar joint set coordinate dataset were previously documented in the cause of the paleostress kinematics analysis of the Anambra Basin lithostratigraphic units (Anigbogu 2013; Okeke et al. 2023; Amogu et al. 2011). In the cause of this research, four normal faults belonging to the Enugu Formation outcropping at the Aguabo pedestrian bridge road cut (along Enugu-Onitsha Road), the Amagu fault domain, Neke Isi-Uzo location and Ozalla/Four corner section aligned with synthetic and antithetic fault arrays along with the Danian horizon of the studied outcrop section at Ikpankwu with diverse fault types and systematic and nonsystematic joints. The fault structural architecture of the formations is confirmed by the displacement of heterolith unit at the mid-section of the Aguabo outcrop section, and the displacement of siltstone, very fined-grained sandstone inter-bed in dominant mudrock facies for the Ozalla/Four Corner, Neke Isi-Uzo and Ikpankwu units.

The Aguabo fault and the Ozalla/Four Corner fault kinematics studies are the prototype fault and joint array utilized for the studied formations' paleostress direction and stress field rotation mechanics.

The Aguabo structural fault style along the Enugu-Onitsha road cut strikes 51° azimuth and dips 53° with dip direction N321°W. The fault design is portrayed by the displacement of the heterolithic unit of the Enugu Formation (Figs. 3.1, 3.2 and Fig. 5.3). The Aguayo fault demonstrates a prototype growth fault design of the Anambra Basin pinpointed by the small-scale rollover structure on the thin fine-grained sandstone bed at the Enugu Formation area. The growth fault architecture was also indicated as the domineering structural style in the subsurface Niger Delta rollover anticline geometry structural analysis (Nwajide and Reijers 1996). The presence of well-stacked late Cretaceous sedimentary rock successions at Enugu and Okigwe area promoted the formation of a growth fault system. The thickness of the sandstone unit on the downthrow block is 10.7 m whereas the thickness on the upthrow block is 8.5 m. A relatively similar thickness of 11.4 m, 10.4 m of the downthrown block and 8.4 m of similar thickness of the upthrown block was recorded by Amogu et al. (2010, 2011) and Anigbogu (2013) respectively. Amogu et al. (2010, 2011) indicated a systematic growth index value of 1.35 for the Aguabo growth fault system. However, the geometry of rollover is a characteristic feature of the hydrocarbon-bearing structures downdip (subsurface) Niger Delta Basin design. This type of fault is a significant sealing fault design that denotes poor fluid migration across the fault plane, synonymous with a potential high-quality reservoir capable of hydrocarbon accumulation and systematic compartmentation.

The Ozalla/Four Corner fault trends 160° (SSE) azimuth, dips 45° and is oriented 85° azimuth. The shale laminar has been sheared as it is bent down at the contact with the siltstone which implies that the hanging wall (siltstone block) moved down relative to the footwall (Fig. 4.6). The throw is 30 cm and the heave, 25 cm. Ozalla fault is seen as an abrupt termination of a meter-thick siltstone bed against a shale unit depicting a series of wedge-like/pinch-out structures of a strati-structural/combination trap model. However, a series of descriptions of sedimentary rock and structural style have been reported on the Ozalla/Four Corner junction fault domain based on the degree of outcrop exposure. The Ozalla fault layer was reported as an abrupt termination of siltstone unit against a dark grey carbonaceous shale (Amogu et al. 2010, 2011). The work emphasized that an intra-formation angular unconformity mirrors the geometry of the Ozalla fracture system which "is interpreted as a normal, low angle fault formed due to slumping and rotation of the material at the right-hand side of the plane". A faulted sandstone juxtapose, characteristic of wedge-like/pinch-out structures illustrative of stratigraphic and structural trap (strati-structural/combination trap) was also documented (Dim et al. 2020; Okeke et al. 2023). The joints at this section of the Enugu Formation form two joint sets of systematic and nonsystematic prototypes which intersects to form a diagonal joint system.

At the time of initiation of these faults, the maximum principal stress $\delta1$ oriented 276° azimuth, the intermediate stress $\delta2$ oriented 10° azimuth and the minimum principal stress $\delta3$ is oriented 179° azimuth. This means that during the formation of the joint, the maximum principal stress was aligned WNW and the minimum principal stress was aligned south with the orientation NNW to NNE.

The principal design of fault movements from the cumulated results of the study domain indicates a dominant extensional tectonic regime of the basin.

Expulsion and tectonic fractures, were documented as the two major genetic types of fractures inherent in structural geology analysis of the earth's crust (Mann 1994). Expulsion fractures are products of petroleum generation, while tectonic fractures are generated from stress build-up forces of the earth's tectonic event. The tectonic regime in this part of the basin is attributed to a regional tectonic model of the Cretaceous Chain and Charcot Fracture Zones due to the lateral extent of the Anambra Basin studied within the Enugu Formation and the Nsukka Formation lithostratigraphic units (Figs. 2.2 and 2.3).

The regional geology report of the Anambra Basin indicated that the basin originated and maintained the status quo as a direct result of the stresses triggered by the earth's tectonic force action along the Chain and Charcot fracture zones that formed the Benue Rift valley (Okeke et al. 2023; Nosike 2023; Umeji 2000; Nwajide 2013, 2022; Murat 1972; Petters 1978; Whiteman 1982; Benkhelil 1989; Ojoh 1990; Nwachukwu 1972; Maluski et al. 1995). A tectonic inversion model from the seismic profile results from the Cretaceous-Paleogene (K/Pg) boundary exposé, indicated that a normal faults system characterizes the Cretaceous-Danian deposit of the Anambra Basin below the Akata Formation. At the same time, the overlying horizon gliding above is a typified reverse fault (Nosike 2023). The phenomenon substantiates the dominant group of Cretaceous Anambra Basin normal faults reported at the Aguabo pedestrian bridge road cut (along Enugu-Onitsha Road), the Amagu fault domain, Neke Isi-Uzo and Ozalla/Four Corner fault along with the Danian Ikpankwu fault system.

However, extensional and compressional tectonic effects were the key principal magmatism folding and faulting generated by the active tectonically induced transcurrent movement along the relatively pervasive fracture zones that extend from the oceanic Charcot Fracture Zone into the existing terrane in the Late Cretaceous. The regional magnetic anomalies and fracture zones substantiated in the Chain and Charcot Fracture Zones triggered series of structural lows that lead to sediment fills with fluctuating marine and continental successions, listric fault and hydrocarbon traps structural style (Nosike 2023). These are evidence of clinoform morphology formed at the continental margin mimicing a series of highland in Enugu area where several kilometers of highlands with a relatively lower dips of an average of 6° was encountered. Santonian tectonic uplift deformation created a high topography and subsiding slopes tapering to the SW and seawards.

However, stress and palaeostress mechanics analysis of the regional fault architecture of the Enugu Formation in the Anambra Basin along with other formations of the basin from slip data on fault systems is beyond the scope of this research. These deformational structure joints, normal fault, growth fault, synthetic and antithetic fault system inherent in the Anambra Basin are consistently similar to the Cenozoic extensional regime with fault system like those of the Anambra Basin. Some previous works in the Cretaceous and Cenozoic basins of the region placed these fault systems within the basin.

5.5 Clinoform Morphology and Structural System of the Anambra Basin

The Anambra Basin and other sedimentary basins in southeastern Nigeria are substantiated by the resultant effect of tectono-sedimentary evolution system of the basin and other adjoining basins. The clinoform and structural fault and joint dynamics of the inland basins in southeastern Nigeria are systematic products and events enacted during the evolution stage and imprint of the Cretaceous tectonic chain and Charcot extensional and the Post Santonian uplift events of the earth's crust. The systematic undulating rolling topography of the studied inland basin is controlled by the sandstone lithology and the underlying bedrock. The underlying bedrocks uplift triggered the > 480 m and 600 m (Okeke et al. 2023) high clinoforms prone to intense erosion and degradation with an estimated > 9000 m clinoform height at the pick of the orogeny/uplift section (Fig. 1.2).

According to Nosike (2023) this Cretaceous highland triggered anchors and eustacy-driven deposition of clinoforms at slopes along with emplacement or activation of diverse normal faults with open and closed architectural design due to displacement downslope. The clinoform mechanics and topography (morphology) of the late Cretaceous Anambra Basin lithostratigraphic units are similar to the modern-day continental shelf margin dynamics of the clinoform morphology depositional pattern that posits the subsiding slopes of the delta as the authentic evolution model of Cretaceous highs (Nosike 2023).

Several evolution model of the Cretaceous Anambra Basin and the Southern Benue Trough recorded in this work were proposed to form by rifting due to the separation of the South American and African Plates in the middle Mesozoic (King 1950; Burke et al. 1972; Nwachukwu 1972). This separation mechanics is the product of triple junction which formed the Atlantic Ocean whereas the failed arm developed the structural axis for the rifting and troughing of the African continent. Two rifting system of divergent pre-rift stages of lower Cretaceous and syn-rift stages of Aptian were reported in the evolution of the basin and other small scale rifting processes (Offodile 1976; Wright 1976; Petters 1978; Benkhelil 1989; Murat 1972). The tectonic effects of the aulacogen (a failed arm of the triple junction) exhibit extensional structure visible within the northern Niger Delta, Anambra Basin and the Benue Trough.

The Cretaceous sediments below the basal unit of the Akata Formation are consistent with hardening from magmatic effect and series of normal faulting systems (Nosike 2023). According to Nosike's work the Cretaceous sedimentary succession of the Anambra Basin is deposited parallelly below the Cenozoic Niger Delta basin deposited from offshore to the exposed sections of the modern Anambra lithofacies units within the basin. However, as the Cenozoic Niger Delta sediments thickened southwards to the offshore direction, the Cretaceous lithofacies thickened and was uplifted northwards. The Santonian deformative tectonic uplift geology events in the late Cretaceous Anambra Basin evolution with the generated geologic highs is regarded as the structural anchor, hallmarked by the deposited clinoforms at slopes, due to eustacy, and activation of diverse normal fault prototypes that is subject to earth's movement downslope. This clinoform occasion is domiciled within the Enugu domain where sediments are deposited along wide subsiding slopes substantiated as a tecteno-sediemnatry event with merged structural and/or sedimentary process imprints eg growth fault, pinchout (Figs. 1.2 and 3.5).

The Abakaliki depression and the Anambra platform south of the Benue Trough are the systematic products of crustal stretching of the lower Albian, within the Gulf of Guinea while the collision between the African and the European plates is the resultant effect of the Santonian shortening stage and the creative folding of the structures. Abakaliki Anticline depression in the southeastern depression domain was triggered by the tectonic inversion, whereas the Anambra platform subsidence activated the development of the Anambra Syncline towards the Northwest direction. The tectonic evolution standard dynamics of the basins and region played a crucial role in the substrata, slopes, subsidence and uplift development of the (basin) region in the northeast-southwest structural trend with deformational dynamical changes. The deposition of the sedimentary successions trajectory of the region follows the structural trend at the margins, leading to the sediment deposition perpendicular to the depocenters (depositional belt).

References

Adesida AA, Reijers TJA, Nwajide CS (1977) Sequence stratigraphic framework of the Niger delta basin. Vienna Austria AAPG Int Conf Exib 81:1359
Amogu DK, Ekwe AC, Onuoha KM (2010) Kinematics of faults and joints at Enugu area of the Anambra basin. J Geol Min Res 2(5):101–113
Amogu DK, Filbrandt J, Ladipo KO, Anowai C, Onuoha K (2011) Seismic interpretation, structural analysis, and fractal study of the greater Ughelli Depobelt, Niger Delta basin, Nigeria. Lead Edge 30(6):640–648
Anigbogu CE (2013) Reservoir quality assessment of the Mamu formation in Enugu and its environs, southeastern Nigeria. MSc. thesis. University of Nigeria Nsukka, 200 pp
Benkhelil J (1989) The origin and evolution of the Cretaceous Benue Trough (Nigeria). J Afr Earth Sc 8:251–282
Bjorlykke K (2010) Petroleum geoscience: from sedimentary environments to rock physics. Springer Science & Business Media, 662 pp
Burke KC, Dessauvagie TF, Whiteman AJ (1972) Geologic history of the Benue valley and adjacent areas. In: Dessauvagie TF, Whiteman AJ (eds) African geology. Ibadan University Press, pp 187–205
Chandra D, Vishal V, Bahadur J, Sen D (2020) A novel approach to identify accessible and inaccessible pores in gas shales using combined low-pressure sorption and SAXS/SANS analysis. Int J Coal Geol 228:103556
Clarke RH, Cleverly RW (1991) Petroleum seepage and post-accumulation migration. Geol Soc London Spec Publ 59(1):265–271
Clarke RH, Grant AI, Macpherson MT, Stevens DG, Stephenson M (1988) Petroleum exploration with BP's airborne laser fluorosensor. Proc Ind Petrol Assoc IPA 88–12(15):387–395
Dim CIP, Okonkwo IA, Anyiam OA, Okeugo CG, Maduewesi CO, Okeke, KK, Umeadi IM (2020) Structural, stratigraphic and combination traps on outcropping lithostratigraphic units of the Anambra Basin, Southeast Nigeria. Petr Coal 62(1)
Doust H, Omatsola E (1990) Niger Delta. In: Edwards JD, Santogrossi PA (eds) Divergent/passive margin basins, vol 48. American Association of Petroleum Geologists Memoir, pp 201–238
England WA, Mackenzie AS, Mann DM, Quigley TM (1987) The movement and entrapment of petroleum fluids in the subsurface. J Geol Soc London 144:327–347
Germeraad JH, Hopping CA, Muller J (1968) Palynology of tertiary sediments from tropical areas. Revue Paleobot Palynol 6:189–343
Hazra B, Chandra D, Lahiri S, Vishal V, Sethi C, Pandey JK (2023) Pore evolution during combustion of distinct thermally mature shales: insights into potential in situ conversion. Energy Fuels
Hindel AD (1997) Petroleum migration pathways and charge concentration: a three-dimensional model. AAPG Bull 81(9):1451–1481
Hunt JM (1979) Petroleum geochemistry and geology. W. H. Freeman, San Francisco, 617 pp

References

King LC (1950) Outline and disruption of Gondwanaland. Geol Mag 87:353–359

Knut B (2015) Petroleum geoscience: from sedimentary environments to rock physics

Loucks RG, Reed RM, Ruppel SC, Jarvie DM (2009) Morphology, genesis, and distribution of nanometer-scale pores in siliceous mudstones of the Mississippian Barnett Shale. J Sediment Res 79:848–861

Magara K (1980) Evidences of primary migration. Am Asso Petrol Geol Bull 64:2108–2117

Maluski H, Coulon C, Popoff MT, Baudin P (1995) 40Ar/39Ar chronology, petrology and geodynamic setting of Mesozoic to early Cenozoic magmatism from the Benue Trough, Nigeria. J Geol Soc 152(2):311–326

Mann U (1994) An integrated approach to the study of primary petroleum migration. In: Parnell J (ed) Geofluids: origin, migration and evolution of fluids in sedimentary basins. Geological Society Special Publication No. 78, pp 233–260

Murat RC (1972) Stratigraphy and paleogeography of the Cretaceous and lower tertiary in southern Nigeria. In: Dessauvagie TFJ, Whiteman AJ (eds) African geology. University of Ibadan Press, Ibadan, pp 251–266

Nosike L (2022) A study of the gas fire at a water well in the caritas university premises, Amorji Nike, Enugu. NAPE Annual International Conference and Exhibition, 20 pp

Nosike L (2023) Impact of structural dynamics on hydrocarbon of the deposits in the enugu axis of the cretaceous Anambra basin. PTDJ, July Volume 13, No. 2

Nwachukwu SO (1972) The tectonic evolution of the Southern portion of the Benue Trough, Nigeria. Geol Mag 109:411–419

Nwajide CS (2013) Geology of Nigeria's sedimentary basins. CSS Bookshops Ltd., pp 347–411

Nwajide CS (2022) Geology of Nigeria's sedimentary basins, 2nd edn. Albishara Educational Publishers, Enugu, p 693

Nwajide CS, Reijers TJA (1996) Geology of the southern Anambra Basin. In: Reijers TJA (ed) Selected chapters on geology. SPDC Warri, pp 133–148

Offodile MA (1976) A review of the geology of the Cretaceous of the Benue valley. In: Kogbe CA (ed) Geology of Nigeria. Elizabethan Publishing Co., Lagos, pp 319–330

Ojoh K (1990) Cretaceous geodynamic evolution of the southern part of the Benue Trough (Nigeria) in the equatorial domain of the south Atlantic: stratigraphy, basin analysis and paleogeography. Bull Centres Rech Explor Prod Elf Aquitaine 14:419–442

Ojo OJ, Akande SO (2009) Sedimentology and depositional environments of the Maastrichtian Patti Formation, SE Bida Basin, Nigeria. Cretaceous Res 30:1415–1425

Okeke KK, Mode A, Anigbogu EC, Umeadi IM, Odu NJ, Maduewesi CO, Ulasi NA (2023) Source rock potential, palynofacies depositional environment synthesis and structural traps of the Enugu Formation Southeastern Region, Nigeria. Arab J Geosci 16(5):1–17

Petters SW (1978) Mid-Cretaceous paleoenvironments and biostratigraphy of the Benue Trough, Nigeria. Geol Soc Am Bull 89:151–154

Schowalter TT (1979) Mechanics of secondary hydrocarbon migration and entrapment. AAPG Bull 63:723–760

Short KC, Stauble AJ (1967) Outline of geology of Niger delta. Am Asso Petrol Geol Bull 51:761–779

Singh DP, Singh V, Singh PK, Hazra B (2021) Source rock properties and pore structural features of distinct thermally mature Permian shales from South Rewa and Jharia basins. India. Saudi Soc Geosci 14(916):1–16

Stacher P (1995) Present understanding of the Niger delta hydrocarbon habitat. In: Oti MN, Postma G (eds) Geology of deltas: Rotterdam. Balkema, A.A., pp 257–267

Tissot BP, Welte DH (1984) Petroleum formation and occurrence, 2nd edn. Springer, Berlin, p 699

Umeji AC (2000) Evolution of the Abakaliki and the Anambra sedimentary basins, southeastern Nigeria. In: A report Submitted to The Shell Petroleum Development Company Ltd, p 155

Ungerer P (1990) State of the art of research in kinetic modelling of oil formation and expulsion. Org Geochem 16:1–26 van der Pluijm BA, Marshak S (2004) Earth structure: an introduction to structural geology and tectonics, 2nd edn. Norton, New York

Whiteman A (1982) Nigeria: its petroleum geology, resources and potential, vol 2. Graham and Trotman, London, p 394

Wright AE (1976) Review of the Origin and Evolution of the Benue Trough in Nigeria. Open University, Milton Keynes, Department of Earth Sciences

Conclusions

- A detailed review of source rock potential of the Enugu Formation at Neke Isi-Uzo, Aguabo pedestrian bridge road cut, Amagu and Ozalla/Four Corner outcrops along with the Danian units of the studied outcrop at Ikpankwu area, Anambra Basin explored palynofacies elements, colour, quality and quantity of the particulate organic matter recorded in the study which suggests the hydrocarbon source potential illustrated as Kerogen Type II/III which are Gas/Oil prone. Interpreted Thermal Alteration Index (TAI) of 2 + to 3 and Vitrinite Reflectance Index (%Ro) values of 0.5% to 1.0% were noted.
- The source rock potential of the study indicates carbonaceous mudrock facies of the Enugu Formation and Danian unit of the region as the source or seal rocks while the siltstones, mixed heteroliths and sandstone units are likely reservoir rocks of the proposed hydrocarbon system.
- The quantity and quality of terrestrial and marine palynofacies constituents in the studied Enugu Formation and the Danian sections highlights a relative abundance of amorphous organic matter, medium and large brown structured phytoclasts, pollen and spores, dinocysts, opaque debris and other marine taxa.
- The synthesized particulate organic matter and sedimentary lithofacies units palaeoenvironment reconstruction exposé favor series of transitional marine environment fluctuations within the inner neritic zone consistent with estuarine, nearshore and fresh water environments during the deposition of the Enugu Formation. An intermittent outer neritic influenced depositional setting is also suggested based on diverse AOM elements, foraminifera test linings and marine dinocysts micro fauna along with relative numbers of land derived plant material palynofacies group. The paleoenvironment system of the studied Danian strata depicts shallow marine environment with deep marine influenced macro environments embedded as products of lower and upper deltaic plains; oscillating from tidal flat, lagoon, tidal bar to nearshore settings with deep marine influences.

- The important structural hydrocarbon traps of the Enugu Formation at the Neke Isi-Uzo area, Aguabo pedestrian bridge road cut, Amagu and Ozalla/Four Corner road cut outcrops along with the Danian sediments of the region outcropping at Ikpankwu locality contain the growth fault, normal fault and anticlinal structures with hanging wall closures, footwall closures, anticlinal closures, horst blocks, half-grabens, and grabens with related rollover structures linked with conjugate fault arrays along with infilled (closed) and open joint structure patterns within faulted and unfaulted sandstone lenes/units.
- Primary and secondary hydrocarbon fluid migration pathway of the study is through sandstones pores and fractures which has been considered as the principal process in the formation of petroleum accumulations in reservoir beds. The external geometry of reservoir rocks is largely determined by the depositional environments (sedimentology), provenance (Composition of sediment), but faulting and diagenesis may define the lateral or vertical extent of a reservoir.
- The natural gas flare and series of oil seepages within the studied formations highlight the described structural styles and causal links required for the hydrocarbon generation, migration pathway and trapping mechanics within the Enugu Formation of the Anambra Basin. The kinematics of faults and joints along with the clinoform morphology system in the studied formations is vital in the hallmark comprehension of the orientation pattern, paleostress architecture of the extensional regional fault along with the rifting and troughing system of the region for effective hydrocarbon accumulation, exploration and exploitation of potential petroleum targets.

The manufacturer's authorised representative in the EU is Springer Nature Customer Service Centre GmbH, Europaplatz 3, 69115 Heidelberg, Germany. If you have any concerns regarding our products, please contact ProductSafety@springernature.com

Printed and bound by CPI Group (UK) Ltd, Croydon, CR0 4YY

26/03/2026

02078952-0015